Framing

A Practical Manual of

APPROVED UP-TO-DATE METHODS OF HOUSE FRAMING AND CONSTRUCTION
TOGETHER WITH TESTED METHODS OF HEAVY TIMBER AND PLANK
FRAMING AS USED IN THE CONSTRUCTION OF BARNS, FACT-
ORIES, STORES, AND PUBLIC BUILDINGS; STRENGTH
OF TIMBERS; AND PRINCIPLES OF ROOF
AND BRIDGE TRUSSES

Edited under the Supervision of

WILLIAM A. RADFORD

PRESIDENT OF THE RADFORD ARCHITECTURAL COMPANY,
EDITOR-IN-CHIEF OF THE "AMERICAN CARPENTER AND BUILDER"
AND THE "CEMENT WORLD"

Assisted by

ALFRED S. JOHNSON, A. M., PH. D.

EDITOR IN CHARGE OF CYCLOPEDIA DEPARTMENT
THE RADFORD ARCHITECTURAL COMPANY

and

BERNARD L. JOHNSON, B. S.

EDITOR OF THE "AMERICAN CARPENTER AND BUILDER"

ILLUSTRATED

Books for Bussines
New York - Hong Kong

Framing:
House Framing, Barn Framing,
Roof Framing

edited by
William A. Radford

ISBN: 0-89499-195-7

Reprinted from the 1909 edition

Books for Business
New York - Hong Kong
http://www.BusinessBooksInternational.com

Table of Contents

Framing

The subject of Framing, taken in its broadest sense, includes pretty nearly the entire structural field. By one of the most common uses of the term, whenever two members are joined or fastened together they are said to be **framed** together. More especially, this applies to the heavy or supporting members of any structure. Thus we speak of the steel framing of a modern "skyscraper." Most framing, however, implies wood construction, as the timber framing of trestle bridges, heavy framing for barns or public buildings, and the framing for houses of various sorts.

To the carpenter especially, and to all others interested in wood in a structural way, this is a most important subject. The framing of a building has been likened to the skeleton of the human body; it is important that it be put together properly and connected up in the right way. The whole stability and success of the edifice depend on the strength and proper arrangement of the supporting frame. Also, when the framing in its various forms and with its various allies has been mastered, the whole structure will be understood.

1

In examining this subject with special reference to practical carpentry construction, framing may naturally be divided under the following heads:

(1) Timber framing for houses;
(2) Barn framing;
(3) Framing of factories, stores, and public buildings;
(4) Miscellaneous framing, including strength of timbers and the principles of truss construction.

The work, accordingly, will be taken up in this order. In some cases, certain subjects of an introductory or explanatory nature will be discussed, although, strictly speaking, they are no part of "framing," and, possibly, are not done by the carpenter. Yet a knowledge of them will add to the carpenter's equipment, and will help him to do his work more intelligently.

Frame Houses

Framing Complete from Foundation to Roof

Taking up the subject of **house framing**, it is to be noted at the outset, that certain very important work must be done before anything like a stable, permanent structure can be erected. The plan must be laid out, building lines determined and fixed, excavations made, and foundations erected.

Building Lines. After the site has been determined exactly—either in accordance with the architect's drawings or determined by the soil conditions, elevation, grade, etc.—it is in order to **stake it out.** This is done by placing stakes outside of each corner and connecting them with cords to guide the workmen in their excavating. It is very important that this be done with great care. Even in small buildings, it should be carefully attended to; while for large structures this work is entrusted to an engineer, who lays out the building lines with transit and level. The lines that have to be located are: the **excavation line** (which is outside of all); the **face line** of the basement wall; and, for masonry construction, the **ashlar line,** which indicates the outside face of the brick or stone wall.

The main rectangle of the plan is laid out

first; and then the supplemental rectangles—as for ells, porches, bays, etc.—are laid out with reference to it, in their proper places.

Squaring a Corner; the 6, 8, and 10 Rule. It is frequently required to square the excavation for a building with a tape line, without the use of other lines and stakes. The method is simple,

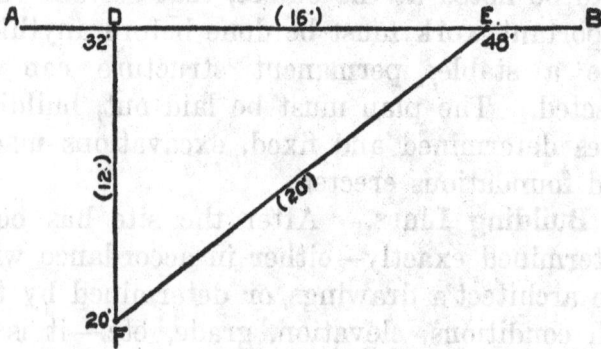

Fig. 1. Squaring Work by the 6, 8, and 10 Rule.

and may be quickly done by three parties, as follows: Run off 24 feet; then the first party should take the end of the tape, and hold it at the 24th foot. The second party should hold the line with thumb and finger at the 16th foot, and the third party in like manner at the 10th foot. Draw till the line is tight, and it will form a right-angled corner true enough for proving up excavation work. The figures given are absolutely correct; but, as a little is liable to be lost at the corners in not being able to hold the tape so as to make sharp bends, this may cause a trifle variation, but the results will be true enough for the purpose stated. The tape will outline a triangle with its sides 6, 8, and 10 feet long; hence the method is called **the 6, 8, and 10 Rule**.

Other figures may be used, provided they are in the same proportion—as 12, 16, and 20. The illustration, Fig. 1, shows the latter figures applied. Suppose we wish to square-cut from the line **AB** at **D**. Measure back on the line from this point 16 feet, as at **E**, and with the end of the tape drawn to 48 and stationed at **E**, the 32-foot mark will be at **D**, and the third point will be at the 20-foot mark at **F**; then **DF** will be at right angles to **AB**.

Another method which is very convenient to use at times is as follows: Draw a line, Fig.

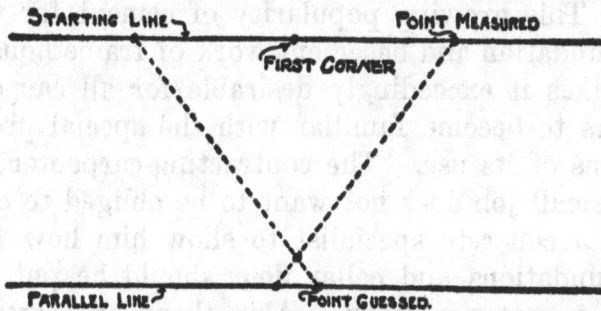

Fig. 2. Method of Squaring Work.

2, parallel to the starting line at opposite side or where there is to be an angle. Now, stick a pin in line at corner stake, and measure a given distance each way on the line—usually about as far as the parallel line is distance away. Now, from one of the points measured, draw a tape to as nearly square across from the corner as can be guessed at, and place a pin there. Then measure the same distance from the other measured point, and stick another pin. Divide distance between these two pins, and you are "squared" across from first corner; and the rest is easy.

Concrete Foundations. Except in a few localities where native stone is to be had very cheaply, all foundation walls are coming to be of concrete. Builders have found that for strength, warmth, and enduring qualities, foundation and basement work in this material is far superior to brick or to wood piles; and for economy and ease in handling, it has an advantage over stone.

This growing popularity of cement for the foundation and basement work of frame houses makes it exceedingly desirable for all carpenters to become familiar with the special problems of its use. The contracting-carpenter on a small job does not want to be obliged to call in a concrete specialist to show him how the foundations and cellar floor should be put in. It is not necessary. Also there are certain problems in connection with the joining of the wood construction onto the concrete that are worthy of attention.

There are a number of types of concrete foundation walls now accepted in general use. Two are illustrated in Fig. 3. They are: first, the entire foundation wall of cement blocks;

CEMENT BLOCK VE-
NEER ON BLOCK
FOUNDATION.

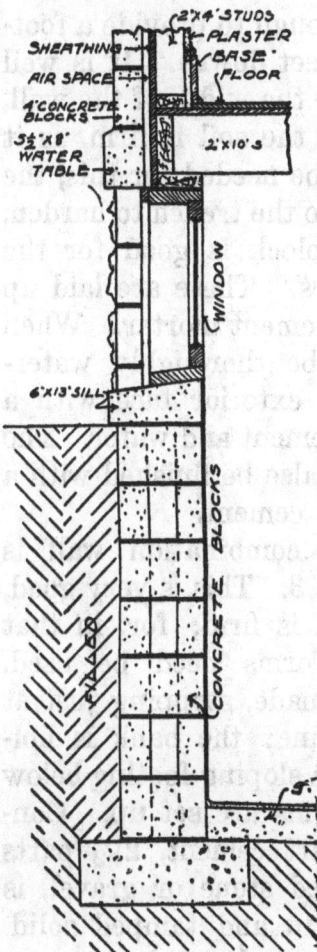

FRAME CONSTRUCT-
ION ON COMBINED
BLOCK & POURED
CONCRETE FOUND'N.

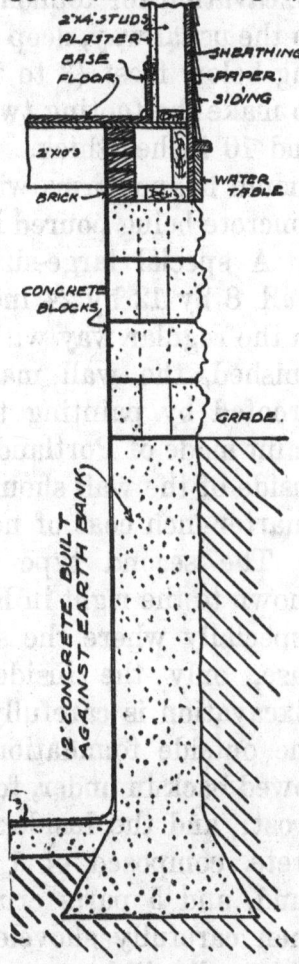

Fig. 3. Two Standard Types of Concrete Foundations for Houses

second, the combination wall (poured concrete
to grade, and blocks or dressed stone above).

A wall of the first kind is shown to the left.
Excavation for foundation of this kind is made
in the usual way, deep enough to provide a footing below frost (3 to 5 feet down). It is well
to make the footing twice the width of the wall,
and 10 inches thick. If the soil is firm, as it
should be, no forms will be needed for this, the
concrete being poured into the trench to harden.

A special large-size block is good for the
wall, 8 by 12 by 24 inches. These are laid up
in the regular way with cement mortar. When
finished, the wall may be thoroughly waterproofed by painting the exterior face with a
paint made of Portland cement and water. The
inside of the wall should also be finished with a
quarter-inch coat of neat cement.

The second type or combination wall is
shown to the right in Fig. 3. This is very good,
especially where the soil is firm; for, in that
case, only the inside forms need be used.
Excavation is carefully made, stopping just at
the outside foundation line; the bank is hollowed back in under, for a sloping footing below
frost; and the inside forms are set up. Concrete, composed of 1 part cement, 2½ parts
sand, and 5 parts crushed stone or gravel, is
then carefully shoveled in and tamped solid.
This wall will be waterproof, dense, impervious
to water, if, before the Portland cement was
used, hydrated lime in the proportion of 1 to

10 was thoroughly mixed through it. When this foundation has hardened sufficiently, the upper wall of blocks or dressed stone is laid up in the regular way.

Fig. 3 shows also two methods of framing for the superstructure—one for an ordinary frame building, standard construction, and the other for a frame building veneered with four-inch-thick concrete blocks. These should be secured to the framework, either with patent anchors or with large spikes driven into the wood with the heads built into the joints.

An **economical foundation wall** sometimes used where the building code prescribes thick walls, is a combination of hollow block and monolithic construction. Its economical features are not confined alone to the saving of concrete, but include the forms also, as scarcely any form is necessary for the footing; and after that the piers are built, requiring but a few forms, which can be used over and over again without resawing or wasting lumber. Fig. 4 shows the arrangement. Piers 6 or 8 feet apart are erected, using a grooved block. Between these piers the curtain walls are placed after the piers become hard.

On small work, where only two or three men are employed, no stop need be made if four piers and three sections of curtain wall forms are used.

The water-table is made after the piers and curtain walls are self-sustaining.

By the use of hollow blocks for piers and monolithic curtain walls, this method of construction is surprisingly rapid and effects a great saving of cost, especially in localities where the hauling adds much to the cost of concrete.

The appearance of this wall is preferable to

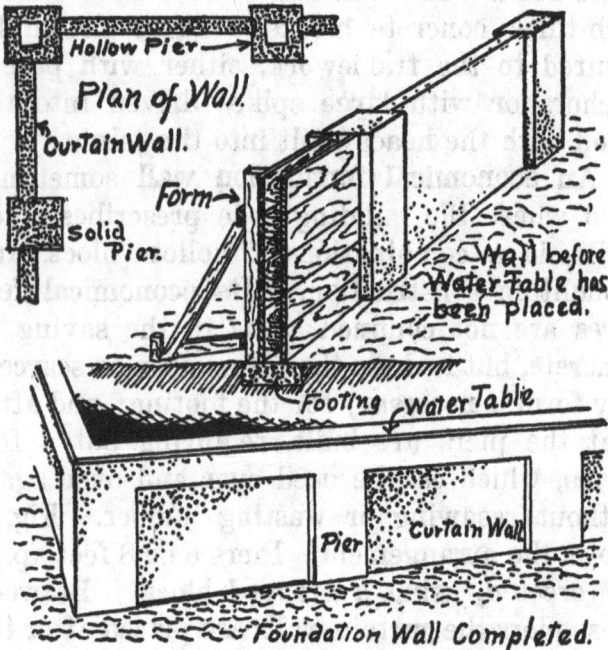

Fig. 4. An Economical Concrete Foundation.

that of the straight plain type. Besides, when building codes class concrete with rubble stone walls in thickness, only the piers need be the thickness required, while the curtain wall is usually acceptable if six inches thick. Walls of this type have been made as light as four inches, and have stood every test.

With this method, a single wagon carries all forms and tools from one job to another; the cost of the forms, made of surface lumber, is about $18.00, while the waste of lumber on a complete form for a dwelling foundation wall 30 by 40 by 7 feet high, for a 12 or 18 inch wall, will be $25, to say nothing of discoloring about $150 worth of good lumber.

By adding about one pound of ultramarine blue to each barrel of cement used for curtain walls, a beautiful effect is obtained, as it gives the piers and water-table a lighter color and more massive appearance.

Cement Cellar Floors. No matter what kind of foundation walls are used, the floor of the up-to-date basement or cellar is of **concrete.** The construction is very similar to that for cement sidewalks. No sub-foundation is, however, necessary as a general thing. Level and pack the earth surface and lay down 5 inches of concrete. Float smooth, giving all sections a slight slope toward some common drain point. When the concrete has become slightly hardened, apply a half-inch top dressing of neat cement, or a rich cement mortar. This dressing should be rounded up in the corners and made continuous with the side wall finish.

To provide for cleaning water, or for any other moisture that may get in at any time, a

tile drain leading outside the basement wall should be provided.

Securing Dry Cellars. In localities where a porous or sandy soil exists to the depth of six or more feet, cellars are usually dry without the use of any preventative to dampness; but where compact soil exists, usually about 80 per cent of all present cellars are more or less subject to dampness, as few have been waterproofed. That concrete, like brick and stone, is a conductor of dampness is known; but that it is more readily adapted to waterproofing only those experienced in waterproofing walls below the grade line have appreciated.

Physicians have long realized that a large amount of sickness is caused by damp cellars, but as waterproofing does not add to the appearance while adding to the cost, it is usually omitted, though medical bills more than make up this additional cost in a few years.

In the illustration, Fig. 5, it must be remembered that the piping shown is for drainage only, and no provisions are shown for sewerage plumbing, which will require separate piping and should never be connected to the drainage sewer nearer to the building than beyond the last trap shown.

A monolithic concrete wall below grade is the cheapest and strongest; and when waterproofed on the outside and on the top with the offset shown, with any positive waterproofing, it will insure dry walls. It, however, causes

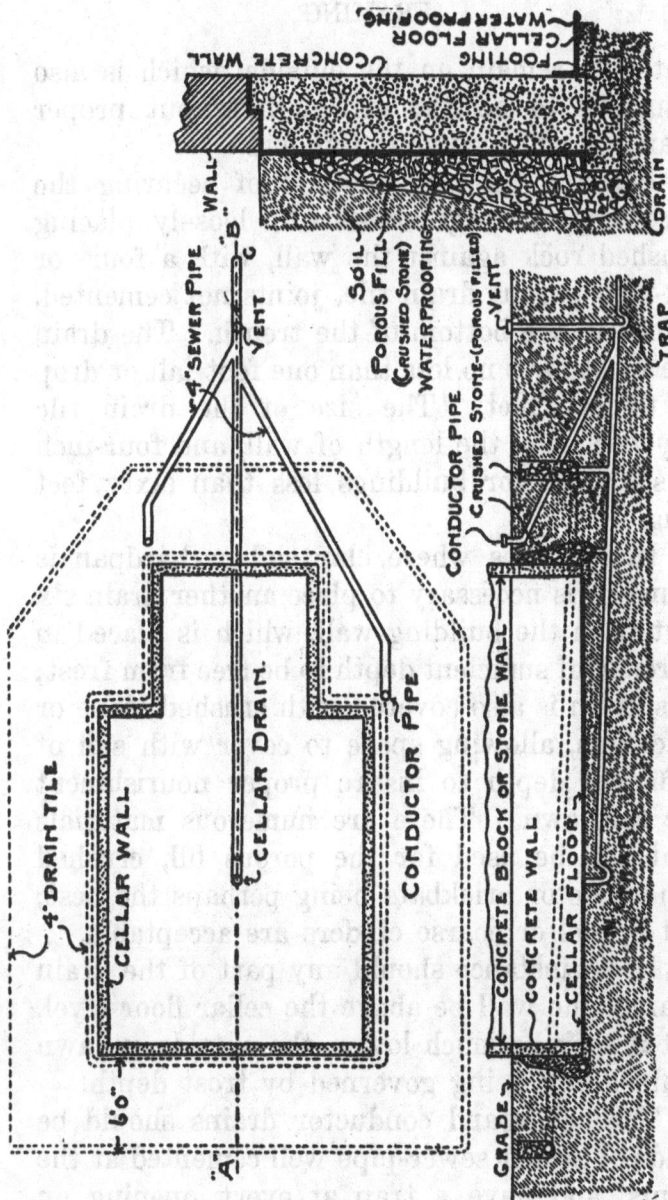

Fig. 5. Waterproofed Cellar with Drainage Piping.

13

water to remain on the outside, which is also injurious to health; and nothing but proper drainage will overcome this evil.

Perhaps the best method of securing the necessary drainage consists in loosely placing crushed rock against the wall, with a four- or six-inch porous drain tile, joints not cemented, placed in the bottom of the trench. The drain tile must have no less than one foot fall or drop in twenty feet. The size of the drain tile depends upon the length of wall, and four-inch is sufficient for buildings less than sixty feet long.

In localities where clay soil or hardpan is found, it is necessary to place another drain six feet from the building wall, which is placed in a trench of sufficient depth to be free from frost; this drain is also covered with crushed stone or brickbats, allowing space to cover with soil of sufficient depth to insure proper nourishment for the lawn. There are numerous materials that can be used for the porous fill, crushed sandstone or brickbats being perhaps the best; but gravel or coarse cinders are acceptable.

In no instance should any part of the drain nearest the wall be above the cellar floor level, but it may be much lower, the outside or lawn drain depth being governed by frost depth.

The cellar and conductor drains should be made of socket sewer-pipe well cemented at the joints, and have a trap at every opening on the inside of the building, and one trap after

all connecting drains have been entered into
the outlet; and this trap must have a vent-pipe
to prevent the formation of noxious gases.
Some contend that the conductor pipe having
an iron pipe from the grade to the eaves of the
roof, makes the best vent possible; but drains
connected with street sewers often carry gases
from the sewers when the traps are not water-
sealed, in which event the conductor would be
an outlet for such gases.

Waterproof Cellar near a Stream. It is frequently necessary to build a cement cellar close
to a stream where water is liable to seep in

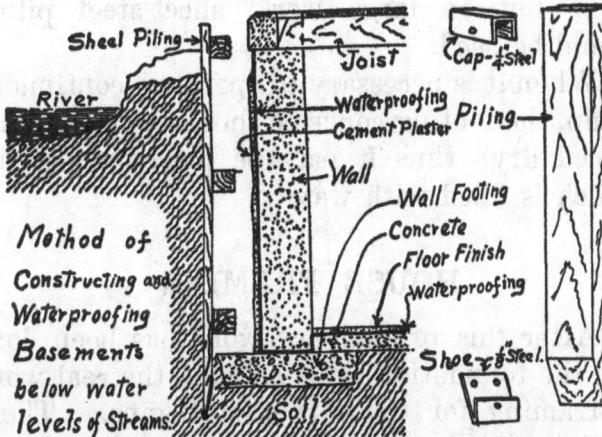

Fig. 6. Waterproofing Cellars below Water Level.

from the bed to the bottom of the cellar when
the water rises during the spring freshet.
Waterproofing for basements so located must
be strong and well made, as it must **resist
pressure.** A good method is to apply the water-

proofing on the outside of the wall, covering same with cement plaster as shown in Fig. 6. On the bottom apply the waterproofing on the concrete body, and cover with the half-inch cement finish (wearing coat). Care must be taken to cover the entire surface to make it absolutely water-tight. On old walls and floors, the waterproofing must be applied on the interior.

The use of sheet piling made of wood, with a sheet-steel shoe for each plank, and a driving cap of sheet steel as shown, saves labor, time, and cost on even the smallest job requiring cribbing; but on large work, sheet-steel piling should be used.

When it is necessary to operate a continuous pump, the footing concrete should be mixed and placed dry; thus it may be placed when the trench is filled with water.

HOUSE FRAMING

After this preliminary work has been done and the foundation walls erected, the real work of framing for the carpenter begins. There are in general use at the present time two distinct types of framing, known respectively as **braced framing** and **balloon framing**. The first is the older and stronger method, and is favored, especially in the Eastern States, for the more expensive houses. The latter has come into much favor during the past fifty years, is decidedly cheaper than the other, and is entirely satisfactory for most residence work.

Balloon Framing. Since the "balloon" system of framing is now most in use, it will be chiefly considered here, with some references to and comparisons with braced framing, as opportunity offers. It may be said in passing, however, that in a full braced frame all the pieces are fastened together with mortise-and-tenon joints; but this is modified in actual practice, spiked joints and corner braces being used.

Sill Construction. The sill is that part of the side walls of a house that rests horizontally immediately upon the foundation or under-pinning, to which it should be securely fastened.

In former times it was required that sills should be of squared, solid timbers of good size, 6 by 8 or at the least 6 by 6 inches. In connection with the other economies introduced with balloon framing, however, several types of **box sills** have come into use, and are thoroughly satisfactory if laid on a good wall foundation. They will not do on posts or piling.

Box Sills. As is always the case where rules governing construction and design are yet in a changing and unsettled state, so-called box sills of various kinds have been used under houses—not altogether satisfactorily. In some cases the studs are set on top of sill or wall-plate, and the floor-joist spiked to the studs, so that when floor is laid out to studs it leaves an opening the height of joist and width of studding from

underneath the building up between siding or
sheathing and plaster, thereby allowing rats,
mice, and cold air free access.

Some carpenters try to remedy this by blocking in between the studs, which, if well done,
will answer the purpose of closing the opening;
but another objection still remains.

The sills, joists, studs, and other rough lumber, are generally right from the saw, and, upon
seasoning, will shrink from ¾ to 1 inch to the

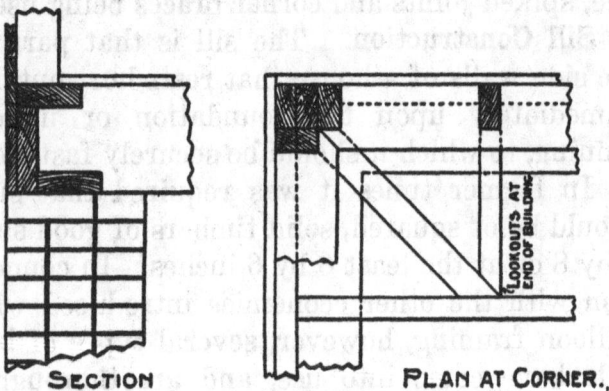

SECTION   PLAN AT CORNER.

Fig. 7. Box Sill Construction.

foot in width, while the shrinkage in length is
scarcely perceptible. So the joists are on top
of sill and nailed to studs, and floor laid out to
studs, and base fitted close to the floor and
nailed to the studs. This looks all right, and
would be if it stayed so; but in a few months
the joist will shrink and take the floor down
with it, leaving the base nailed to the studs, and
a crack from ½ to ¾ inch under the base.

A much better plan is shown in Fig. 7. Lay a 2 by 8 on foundation wall flatwise, and size the ends of floor-joist on top to receive a 2 by 6 around the entire building, and set studs on this directly over joist. Then, when floor is laid, it will lap 2 inches on the 2 by 6 if 2 by 4 stud is used, or 1 inch if 2 by 5 stud is used, thereby effectively stopping all openings

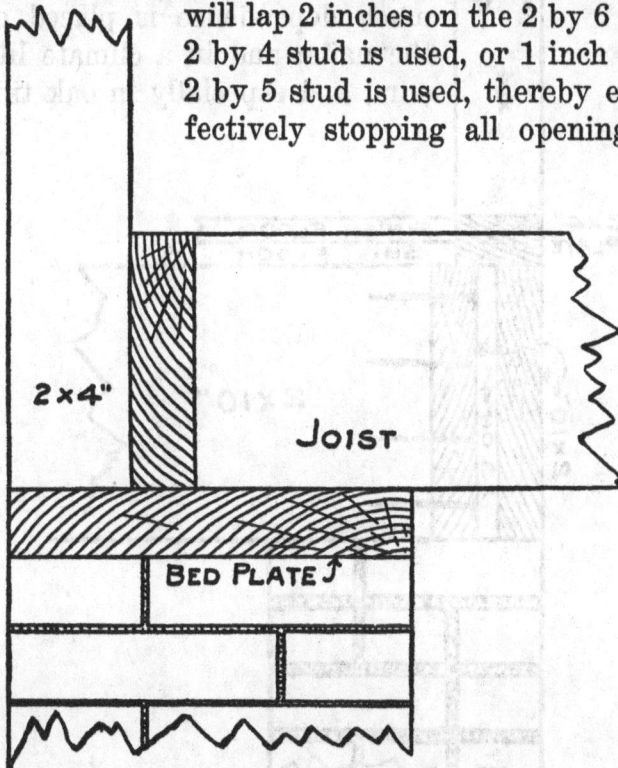

2×4"

JOIST

BED PLATE

Fig. 8. Box Sill Construction.

for draughts of vermin. But this is not all; when the floor-joists shrink, the studs go down with them, thus keeping base, floors, doors, etc., in their original position relative to one another.

In case posts are used instead of solid wall,
it is then necessary to use a sill instead of wall-
plate, and joist should be gained down level with
top of sill. With a box sill or built-up sill of any
kind for a post foundation, too
much dependence is placed on
the nails; and in a climate like
ours, and especially in oak tim-

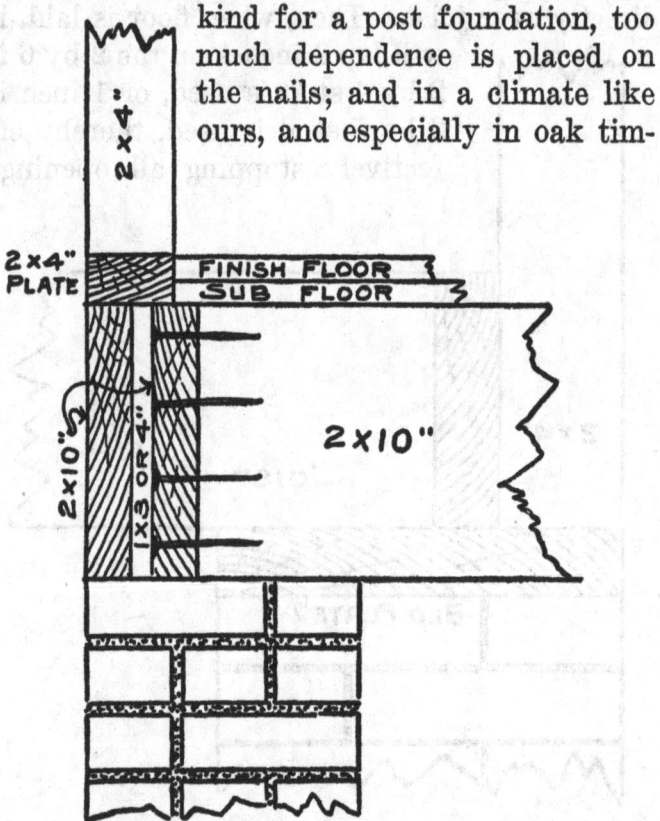

Fig. 9. A Built-Up Sill.

ber, the nails, in the course of a few years,
become very weak and brittle from rust.

Another box sill that is claimed to be strong,
warm, and "rat-proof" is shown in Fig. 8.

First there is a wall-plate or bed-plate, say of
2 by 8 inches, and an upright the same width
as joist, thus allowing the studding to rest on
the bed-plate and be spiked to the upright.

With a sill made this way, you
get the full strength of the
studding; there is less work
cutting same; and it is easier

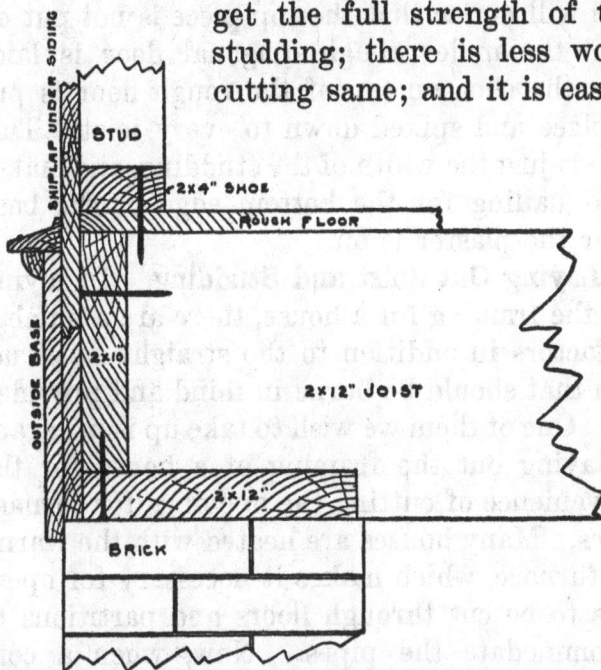

Fig. 10. Box Sill Construction.

to lay the floor, as there is no studding to cut
around.

A **built-up sill** that has a number of good
points is illustrated in Fig. 9. This construction makes a very strong sill for a solid wall
foundation, and makes the walls perfectly rat-
proof and also draught-proof. In case of using

a sub-floor, the sill makes a good support for
the ends of the sub-floor.

A box sill of still another type is detailed
in Fig. 10. With it the building rests on a
twelve-inch bearing on the foundation wall.
You will notice that the top piece is not put on
until the under rough diagonal floor is laid;
then the shoe on top of the rough floor is put
in place and spiked down to every joist. This
shoe is just the width of the studding, and makes
good nailing for the bottom edge of the base
after the plaster is on.

Laying Out Joist and Studding. In laying
out the framing for a house, there are a number
of factors in addition to the straight construction that should be borne in mind and provided
for. One of them we wish to take up is in regard
to laying out the framing of a house for the
convenience of cutting the openings for furnace
pipes. Many houses are heated with the warm-
air furnace, which makes it necessary for openings to be cut through floors and partitions to
accommodate the pipes. Now, when a contractor has a house to build that is to be heated
with a furnace, the first thing he should do is
to properly locate all the registers on the plans,
if they have not been located by the architect.
Some architects leave the location of the registers to the furnace contractor, but a competent
architect can locate the pipes and outlets for
furnace heating to just as good advantage as
the furnace contractor, and in making plans the

architect should locate the place for registers
and everything pertaining to the heating of the
house.

When the carpenter lays out the joists and
partitions, he should see that the joists on the

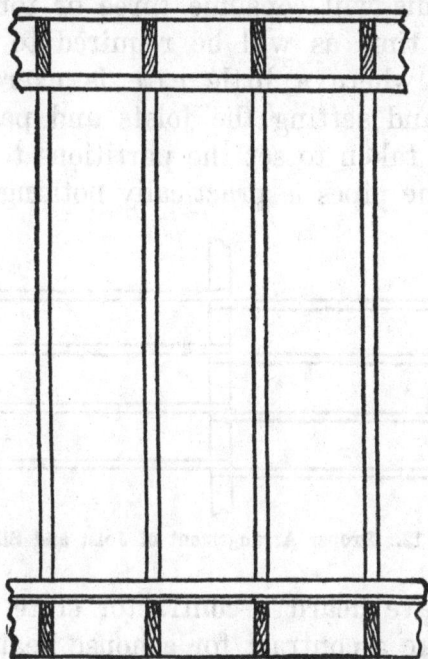

Fig. 11. Proper Arrangement of Joist and Studding.

floor above are directly over the lower ones, and
that the studding are directly under the joists
and in line with them, as in Fig. 11. When
this is done, there is no trouble in cutting the
opening for the pipes and registers. On the
other hand, if this is not done, and the studding
and joists are put in without any regard to

furnace pipes, then the contractor will find his
troubles begin to multiply as soon as he begins
to cut for the furnace pipes. He will find studding that must be moved; he will find joists
that must be moved or cut into and headers put
in; all this will consume three or four times
as much time as will be required to cut the
openings where a little care is exercised in
spacing and setting the joists and partitions.
The time taken to set the partitions to accommodate the pipes is practically nothing.

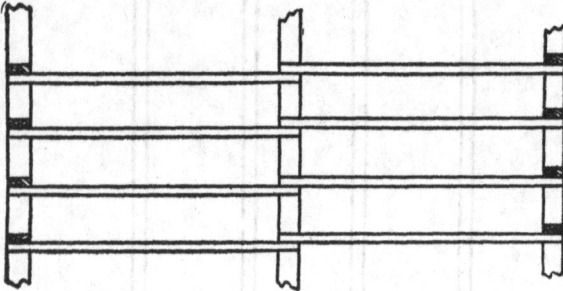

Fig. 12. Proper Arrangement of Joist and Sills.

We have heard a contractor state that he
would take a contract for a house heated with
hot water or steam for fifty dollars less than he
would if heated with a warm-air furnace,
because of the immense amount of cutting the
furnace work requires. The above statement is
not well founded; a contractor with such an
opinion surely does not properly lay out the
framing of the house, if such has been his experience. If the house has been properly laid out,
the cutting for the furnace pipes should not cost

over five dollars. No man should be over one day cutting for furnace pipes in the average residence, and in some cases half this time would be plenty. The way to cut for furnace pipes is to have every pipe located, and to have one man, skilled in this kind of work, do the cutting; then, when he starts to cut, let him keep at it till the job is completed.

It is a great loss of time for a contractor to take men haphazard to do this work, calling a man away from his usual work at various times to cut for the furnace man as he happens to want it. All the cutting should be done at one time, in advance of the furnace man, by some-one who understands it. Working here and there, and cutting for furnace pipes by piece-meal, does not pay.

If a contractor has the cutting to do for steam or hot-water heating, he finds that there is no great amount of difference between cutting for steam or hot-water heating and in cutting for the furnace, if some care is taken in setting joists and partitions.

Fig. 12 shows how studding and joists should be laid out on sills and girders where they lap by one another, to keep them in line across the building. The studding on one sill are set the thickness of the studding to one side of the studding on the opposite sill. Then, when the joists are placed, they can be lapped on the girder in the center, and each tier of joists will line up square across the building. How often

we see joists put in lapped on a girder or partition, and scarcely a joist in the entire floor is
square with the building. This frequently
makes it bad in setting partitions that run parallel with the joists, and sometimes makes extra
cutting necessary.

It pays to be a little particular and exact
in laying out work. The little extra time it
takes is a mere nothing, while its advantages
are very great, and are apparent everywhere
about a building laid out by a competent man
who looks ahead a little in his work. The old
saying, "Never cross a bridge till you come to
it," may be handy for people to say at times,
but be sure the bridge is there when you want
it; it saves trouble.

Wall Framing. Fig. 13 may be taken as a
typical exterior wall section, sill to plate, in the
ordinary cottage or small house. Studs 2 by 4,
set about 16 inches on centers, are stood on the
wall-plate, to which they are toe-nailed. They
are also securely spiked to the floor-joists.

At the ceiling line, a **ribbon** or **ledger board**,
usually 1 by 5, is notched into the studding
and nailed, to support the ceiling joist. At the
top a 2 by 4 **plate** is applied, which supports
the roof rafters through the bearing known as
the **seat cut**.

For cottages, the ceiling joists frequently
project out and meet the rafter ends, thus making an easily constructed box cornice.

Sheathing. We often hear it said that to

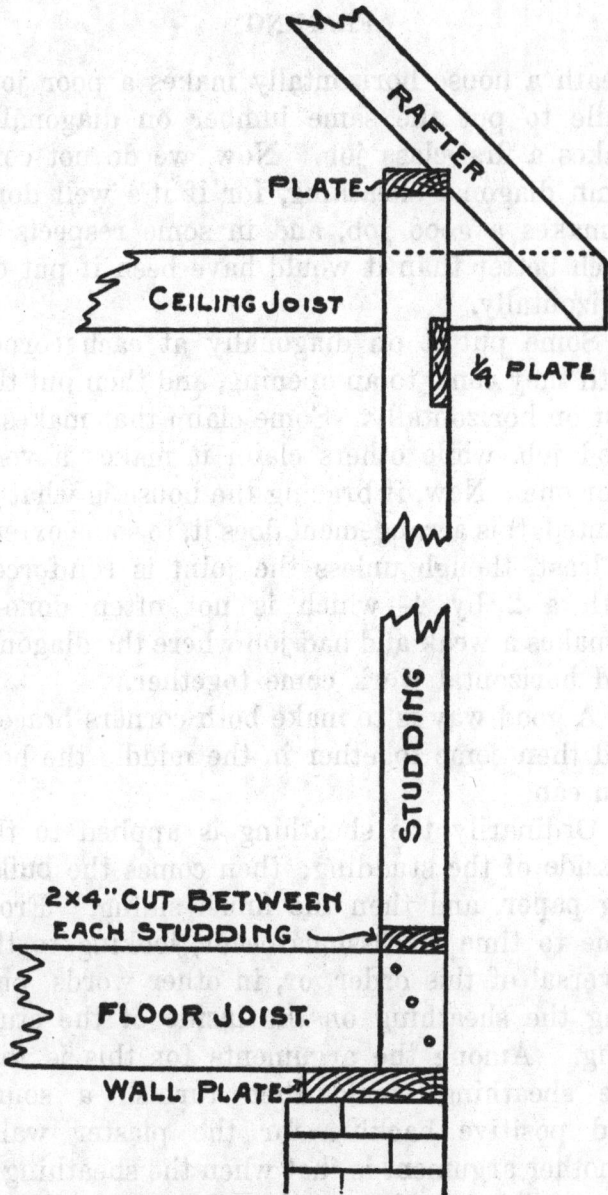

Fig. 13. Wall Construction, Sill to Plate, in Small House.

sheath a house horizontally makes a poor job,
while to put the same lumber on diagonally
makes a first-class job. Now, we do not con-
demn diagonal sheathing, for if it's well done
it makes a good job, and in some respects is
much better than it would have been if put on
horizontally.

Some put it on diagonally at each corner
until they come to an opening, and then put the
rest on horizontally. Some claim that makes a
good job, while others claim it makes a very
poor one. Now, if bracing the house is what is
wanted, this arrangement does it, to some extent
at least, though unless the joint is reinforced
with a 2 by 4—which is not often done—
it makes a weak and bad job where the diagonal
and horizontal work come together.

A good way is to make both corners braced,
and then come together in the middle the best
you can.

Ordinarily the sheathing is applied to the
outside of the studding; then comes the **building paper,** and then the finish siding. From
time to time discussions occur, looking to the
reversal of this order, or, in other words, putting the sheathing on the **inside** of the studding. Among the arguments for this is that
the sheathing would then furnish a sound
and positive backing for the plaster walls.
Another argument is that when the sheathing is
put on the outside and then the weatherboard,
there is some tendency for the moisture to get

in between the two and cause decay; so it was thought better to have the sheathing inside, and the weatherboarding simply as an outside protection on the outside of the studding.

Sound-Proof Walls. Every builder is at some time or other interested more or less in what might be termed **noise-proof** or **sound-proof** walls and floors. It is always desirable to have floors and partition walls as non-productive and as non-conductive of sound as practicable; and there are several different kinds of composition and methods of construction resorted to for this purpose. It is, however, a bigger task than one might think to make an absolutely noise-proof wall. It is said that what is indorsed by Prof. S. I. Franz as the one noise-proof room is a room about 8 feet square and high, on the top floor of the University of Utrecht. Its walls are about 11 inches thick. From the inside, these are made up of successive layers of horse-hair felt, porous stone, dead air, wood partition, ground cork composition, and a plastered surface. The ceiling, though somewhat simpler made, has similar layers. The boards of the floor were sawed, and the joints filled with lead to stop vibration; a layer of lead was then covered over all, to the thickness of more than an inch; and over this, in turn, is used a carpet nearly half an inch in thickness.

This only goes to show to what lengths it is

necessary to go to secure absolutely sound-proof walls.

A method sufficient for all practical purposes, and one that is very often used as a cheap way to deafen the center wall in a double house, so that the occupants of one side cannot hear the other, is shown in Fig. 14. Set a double row of studding, as shown; they are of 2 by 4 inch stuff, set in the usual way, but set staggered, so that the face lines will be 6 inches apart. This will leave a space of 2 inches between the studs and the plastering. Then, on

Fig. 14. An Inexpensive Sound-Proof Wall.

the inner edges of the studs, heavy felt paper or hair insulator quilt should be stretched, and made secure by nailing a lath over the stud, as shown in the section. Two by six-inch plate can be used at the top and bottom. The floors should be deafened, and this can be done very satisfactorily by putting down a rough floor of shiplap, and after all rough work is done, cover this with felt or hair cloth and lay the finished floor. As to back-plastering the wall to deaden the sound, it is not as effective as the above method.

Substitute for Back Plaster. Tar paper

may be used to very good advantage instead of back-plastering a house, provided it is tough enough. The ordinary tar paper, such as is used for covering the sheathing, is generally too soft and is very liable to puncture while the work is being carried on. The accompanying sketch, Fig. 15, shows two ways of doing the work. In the first, by using 32-inch paper and putting it on vertically, it will cover two spaces. All laps should be on solid bearing. Then strip

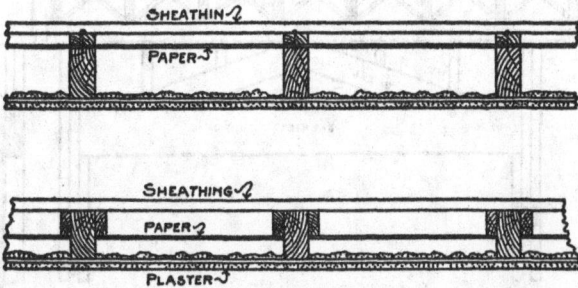

Fig. 15. Wall Sections, Showing Use of Paper Backing.

with ⅞-inch pieces, and put on the sheathing, which should also be covered with paper, and sided in the usual way. Or it may be done as shown in the second sketch in the figure. This, however, requires more cutting and fitting of the paper. The former is preferable in many cases, on account of giving a wider fall at window jambs. The object in this construction is to give as nearly a dead-air space as possible. Therefore every part should be made thoroughly tight, or the object sought will be of no avail.

Double, Trussed Partition. In framing the
wide door opening and double partition for a
large sliding door, the first thing to be considered is the foundation on which to rest the
jambs at the sides of the door. If it is not
convenient to have a partition under the door,
the joist should be doubled, and especially so
if the joist above the door rest or break over
the same. If there is great weight there, the

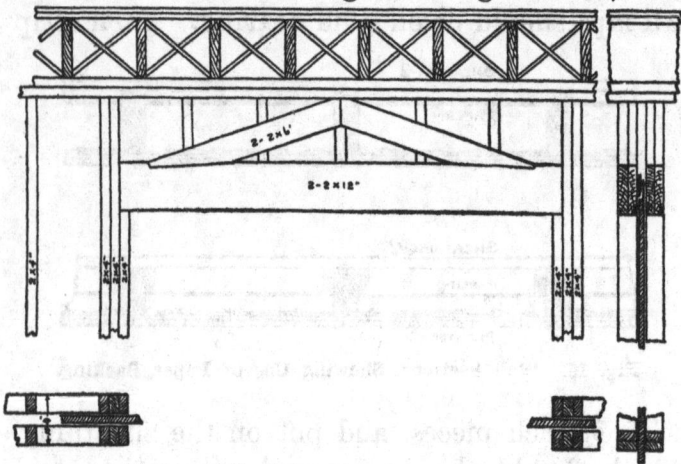

Fig. 16. Double Trussed Partition for Sliding Doors.

joist should be doubled under both partition
walls of the sliding door. Fig. 16 shows good
construction where the joists run at right angles
with the door opening. The truss may be
omitted where the joists run parallel to the
door; but it is a good idea to put in the double
joist at the head of the opening, as it furnishes
an excellent bearing on which to fasten the
track.

Braced Partitions. Quite recently the writer noted for the second time only, in a somewhat wide experience of construction, a very curious and mistaken way of putting braces in a partition. The carpenter had placed the head and sill in position, and then cut and nailed up all the studs in position before the braces. The latter were cut in between the studs, as shown in Fig. 17 (left sketch); and while they did their duty, perhaps, in preventing to some extent the racking of the partition, yet they were not

Fig. 17. Bracing a Partition.

nearly so effective as if they had been fitted in first in one piece, and the studs cut in after, as shown in the right-hand sketch in Fig. 17.

By putting the braces in first, not only is their full strength obtained, but the frame of the partition can be properly squared by measuring or testing the diagonals with a rod in the usual way, and adjusting if necessary—an almost impossible thing to do if the studs are all spiked into position first. Numerous small details of this sort mark the difference between the thoughtful craftsman and the bungler.

Such mistakes come from carelessness or lack of knowledge.

Cornice Construction

A good form of cornice that is very popular, especially for cottages, is shown in detail in Fig. 18. It is what is known as a **boxed cornice.**

Fig. 18. Cornice Construction.

It represents a very easy method of constructing cornice and gutter where the bottom of the roof is concaved. Many will put on the rafters to form the concave portion of the roof, but this is not at all necessary. The concave portion can be made much easier and quicker by cutting in 2 by 3 pieces as shown, and placing the sheathing boards so that the center of the boards will break over the joints of the 2 by 3 where it connects with the rafters and lookouts. Then, when the shingles are put on the roof, they will have just as nice a curve as can be made in any way and at much less cost.

It takes much less time to put in these 2 by 3 pieces than to work out the circular pieces, and

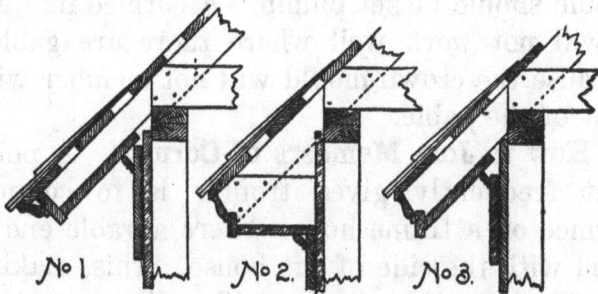

Fig. 19. Cornice Construction.

then the trouble with hips is much less with the straight pieces than it is where hips have to be worked to the circular form. The straight pieces for the hips have to be a little longer than those on the common rafter, but this is very easily found in more ways than one. A little sketch will show it, or it can be figured out mathematically.

Other good cornices are shown in Fig. 19. As illustrated, they are **box cornices**. No. 1 and No. 3 could be used very easily as **open**

cornices—now so popular for bungalow construction—by simply omitting the plancher boards and moulding, the roof sheathing boards at the eaves having been matched and the rafter ends dressed and fashioned as desired.

Where two roofs come together with a valley, it is often a question just how the rafter ends should be cut, square or plumb with the building. If it is a cornice like No. 1, it is proper to cut the ends at the rafters square; but if it is a box cornice like No. 2, then the crown-mould should be set plumb. A cornice like No. 3 will not work well where there are gables, because the crown-mould will not member with that of the gable.

How to Join Members of Cornice. A point that frequently gives trouble is to connect cornice on a frame house where a gable end is flush with the side of the house. This, making the fascia mould member with a like mould on the raking cornice, cannot be done, because the mould on the fascia of the gable is resting on a plumb backing, while that on the raking cornice is at right angles to the pitch of the roof. Consequently, they cannot directly member without making a return connecting the two, as in Fig. 20. This can be very small, just enough to fill the inverted V-shaped gap, where the top edge of the two are on a line, or it may be got over by letting the gable extend a few inches so as purposely to make the return longer and thereby destroy what otherwise may seem to be a blunder on the part of the workmen.

There is another point in connection with a roof of this kind that we wish to call attention to, and that is the cut on the plancher of the gable to member with that of the raking cornice. We have seen carpenters who could readily frame a hip and valley roof, but when they came to make the above cuts, were puzzled to know how to apply the square.

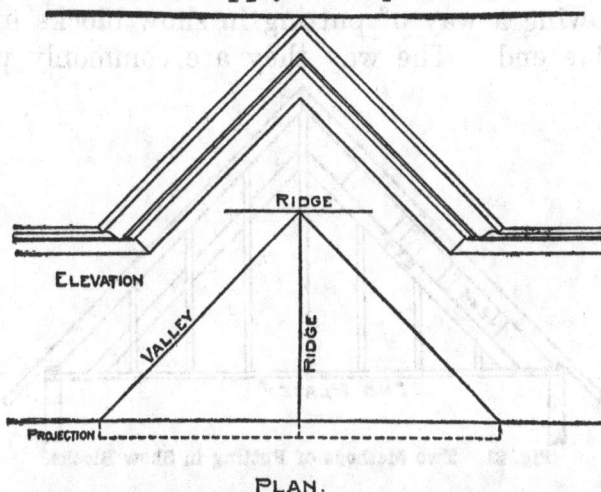

Fig. 20. Gable Flush with Side of House.

The trouble is in this, as in most all other framing problems: they did not stop to think. We make haste sometimes by going slow, and this is one of the times.

Now, let us stop and think. To begin with, the plancher of the gable lies in exactly the same position as the jack rafter. Consequently,

the figures that give the plumb and side cuts of
the jack will give the cuts for the plancher, but
are reversed; that is, the plumb cut becomes
the edge cut, and the side cut becomes the face
cut across the board. As for the cuts of the
raking plancher, it lies in exactly the same position as the roof board just above it, consequently
the same figures that are used for the roof
boards will give the cuts for the plancher.

Putting in Show Blocks. Fig. 21 is a sketch
showing a way of putting in show blocks in a
gable end. The way they are commonly put

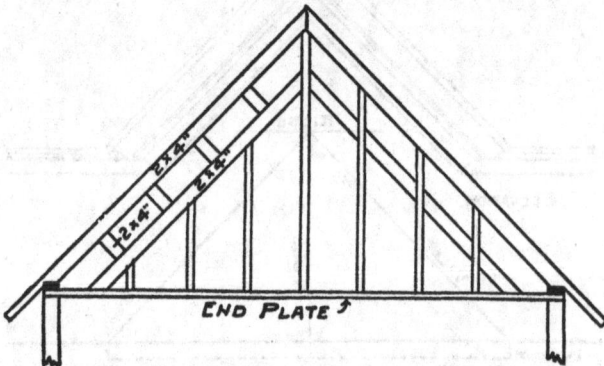

Fig. 21. Two Methods of Putting in Show Blocks.

in is always more or less of an annoyance. In
the sketch, both the common and approved ways
are shown. In the common way, the blocks are
pieces of 2 by 4, cut in between the gable studs,
which, if not extremely well nailed, will become
loosened in nailing on the siding. In the other
way, the blocks are in one continuous piece,
fastened in the center to the gable stud just
low enough to receive the siding and cornice.

While this plan may be old to some, it may be
new to many. It is often some of the simple
things that are of much importance.

Framing for Windows

One of the most important parts of house
framing is the construction connected with the
window openings. This is a comparatively
simple matter, yet it should be done with care
to insure against leaks. One of the best ways

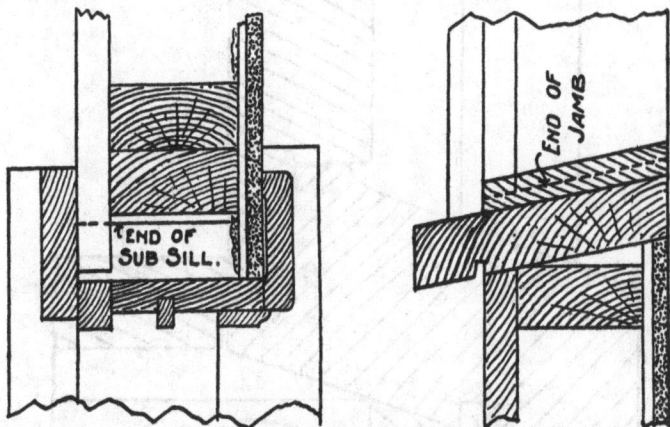

Fig. 22. Simple Window Construction.

to frame to prevent leaks is to gain the jamb
into the sub-sill, letting the end of this sill project same as for window sill, and only notch
out enough of the back corners to fit nicely in
opening for the window as shown by the sectional drawing, Fig. 22. The joints should be
set in white lead, and well painted on the
outside.

Single-Sash Windows. How to make a window-frame for the attic, or a single-sash frame for any place, which will admit of hanging so that the sash can be opened and at the same time keep out the snow and rain in stormy

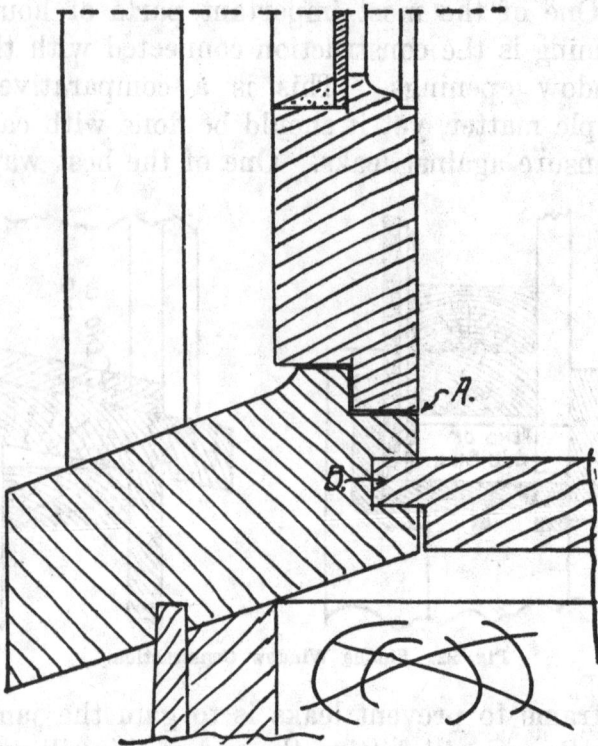

Fig. 23. Framing for Single-Sash Windows.

weather, is a problem that has caused the carpenter more or less study for years.

We have found no better way to accomplish the work than the arrangement shown in Fig. 23.

The sill is made with a lip and then rabbeted, as shown at **A**. The sash is also rabbeted to fit the sill. The sill is plowed at **B**; and the stool rabbeted to fit into the sill. This allows the sash to be hung at the top; and when closed over the rabbeted sill with the lip, it prevents rain and snow from beating in under the sash. If the sash is tightly closed, it is just about storm-proof—so nearly so that not enough rain or snow will get through to do any damage.

With the ordinary window sill, it is impossible to hang a single sash either at the top or on the side, on account of the bevel on the sill; and then the stool is in the way. Our experience has been that almost any attempt to hang a single sash in the frame such as ordinarily made, results in a very unsatisfactory job. If the window happens to be in some place much exposed, it will be found to be a great annoyance on account of leaks.

A frame constructed with the sill and sash rabbeted as shown in the sketch is as near storm-proof as it is possible to get and have the sash hung so that it can be opened readily.

Outward-Opening Casements. Fig. 24 illustrates the construction of an ordinary casement window opening out. The construction is quite simple, and is of the type commonly used in well-constructed frame buildings of the medium class.

The wall is constructed of 2 by 4-inch studs placed 16 inches on centers and doubled about sides, head, and sill of each window opening.

The exterior of the wall is covered with matched sheathing boards, laid diagonally or horizontally, preferably the former way, and well nailed to every stud. Over this sheathing is placed a heavy building paper well lapped and tacked, and carried under outside architraves of windows. Over the sheathing paper the shingles, clapboards, or other final enclosing material are placed.

The course of shingles over window-caps and in similar places is given the required cant by means of a cant strip **K**, tacked to the sheathing boards. The shingles at jambs of windows butt against the outside architraves of same, which architraves should, for this reason, be at least one and one-quarter inches thick. The course of shingles under the window-sills are fitted up in a groove on the under side of the sill. Shingles should be well nailed with two galvanized nails to each shingle, and any shingle over six inches should be split.

At the top of the figure is a vertical section taken through the head of the window. The head of the frame is rebated, and extends from the outside architrave to the trim. The window-cap is flashed with tin as shown, extending up under shingles about 6 inches. Copper is usually used for this flashing in the better class of cottages.

The trim is worked out of 7/8-inch material, is blocked at the back (**B**), and has a 7/8-inch

back-band. The back-band has a feathered
edge at **A**, which is planed off to fit the uneven-

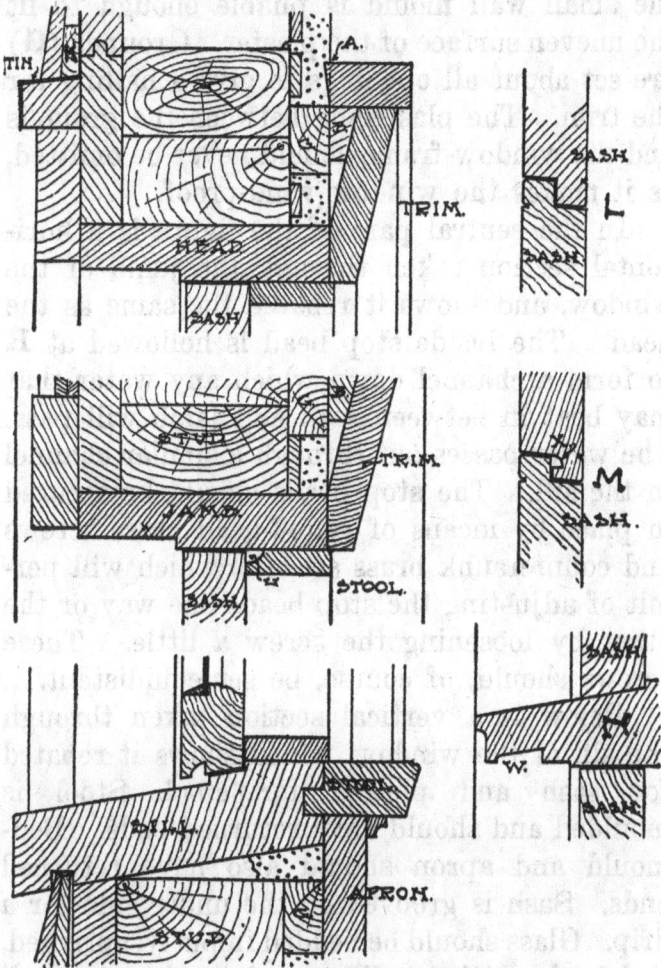

Fig. 24. Framing for Outward-Opening Casements.

ness of the plaster. Frequently a small mould
is provided in the angle formed by the inter-

section of the back-band with the plaster; and in such cases the feathered edge is omitted, as the small wall mould is pliable enough to fit the uneven surface of the plaster. Grounds (**G**) are set about all openings to give a nailing for the trim. The plastering between the grounds and the window-frame should never be omitted, as it makes the window wind-proof.

In the central part of the figure is a horizontal section taken through the jamb of the window, and shows it rebated the same as the head. The inside stop bead is hollowed at **H** to form a channel down which any water that may beat in between sash and jamb will pass. The water passes out through a similar channel in the sill. The stop beads should be secured in place by means of round-head brass screws and countersunk brass sockets, which will permit of adjusting the stop beads one way or the other by loosening the screw a little. These screws should, of course, be set equidistant.

Below is a vertical section taken through the sill of the window, which shows it rebated for sash and plowed for stool. Stool is moulded and should have returned ends. Bed-mould and apron should also have returned ends. Sash is grooved on the under side for a drip. Glass should be bedded in putty, sprigged, and back-puttied. Where plate glass is used, wood beads should be used to hold the glass in position, and putty should be used as a bed for the glass.

Where it is desired to have sashes light in appearance—that is, using as little wood as possible—the stiles and rails are worked in cherry or other hardwood.

L is a section showing the construction of the meeting stiles of casements opening in two leaves. The joint is rebated one-half inch, and edges are beaded as shown.

A somewhat better form of construction is shown at **M**, which shows the sashes rebated and beaded as above described, but with an additional groove at **X** which serves as a conductor for any water that may beat in at the joint, discharging it on the sill outside of the sash.

N is a section showing the construction of a transom bar. A water nose is formed on the projecting part of bar by cutting a hollow, as at **W**.

Another outward-opening casement window is illustrated in detail in Fig. 25, showing one of the best methods of making the frame and sashes, and incidentally showing how the outside architrave may be omitted, which feature is frequently considered desirable from the standpoint of appearance.

The frame wall is constructed in the usual manner of 2 by 4-inch studs, placed 16 inches on centers, and doubled for jambs, heads, and sills of openings. The studs are covered on the outside with matched sheathing boards, heavy building paper, and shingles. The inside of the wall is wood-lathed and plastered. Plaster,

except the white finish, should be carried behind
all wood base, wainscot, trim, dressers, etc.

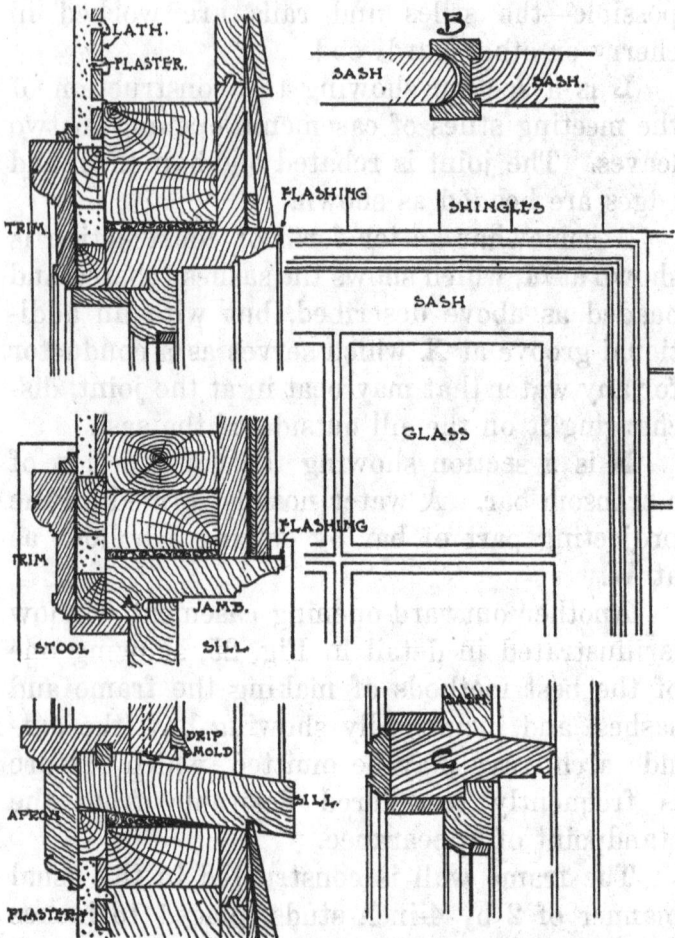

Fig. 25.　Framing for Outward-Opening Casements.

The frame is made from 2-inch stock, rebated
and moulded as shown, and projects beyond the
outside surface of the sheathing boards not less

than one and three-eighths inches, so that the
shingles or other exterior wall covering may
butt against it. This manner of allowing the
frame to project to take the shingles, does away
with the necessity for an outside architrave.
This is considered a desirable feature when it
is important that as little wood as possible shall
show about the window. The head of the window is made water-tight by flashing with tin or
copper in the manner shown. The flashing is
carried up behind shingles at least 4 inches, and
down over top of frame, so that the metal projects sufficiently to form a drip, which prevents
water from trickling down under head of frame.
In the better class of work the joints of the
shingles with the jambs are also flashed in the
manner shown.

After the frame is set, the spaces about the
frame should be made weather-proof by calking
with oakum and plastering over, or by filling up
the interstices with plastering mortar.

The sashes are rebated, and are fitted with
wood beads for holding the glass in place, instead of using putty. The glass, however, is
bedded in putty before the wood beads are af-
fixed. A drip moulding is let into the lower rail
of sash to catch any water drippings and prevent them from beating in under the sash.

Jambs of frame have a groove cut in the rebate as shown at **A**. This groove catches any
water which may beat in between sash and jamb,
and conveys it downward, as indicated by the

dotted lines, and discharges it on the sill as indicated by the arrow.

At the top of the figure is a vertical section taken through the head of the window. The cant strip shown is required over the head to give the first course of shingles the proper tilt. The trim is of 1⅝ by 5-inch moulded and hollow-backed material, mitered at angles, put together with slip tongues, and glued. A small wall mould follows about all trim.

The middle part of the figure to the left is a horizontal section taken through the jamb of the window, and shows the trim and inside stop bead finishing on a moulded stool.

Below is a vertical section taken through the sill, which shows the joint of stool with sill put together with a slip tongue and glued. The sill has an undercut into which the shingles are fitted.

B is a horizontal section through the meeting stiles of casements in two leaves. This makes a tight joint, but requires that both leaves be opened and closed together. The crescent-shaped space between the two stiles is required for the play of the sashes when thrown forward, owing to the hinges turning on a center outside the sashes.

C is a vertical section taken through the transom bar. The projecting portion is pitched to throw off water, and the under side is grooved to form a drip. The upper or transom sash is stationary, and joints are put together with white lead. The inside of the transom bar is moulded to correspond and miter with a portion of the trim.

Inward-Opening Casements. Fig. 26 shows the construction of inward-opening casements in frame walls. The construction is very good for the reason that all precautions are taken to make the window proof against rain and wind.

The sash and frame are rebated at the jambs and sill, and a small mould is tongued into the jambs and head outside of the sash in the manner shown. This mould is undercut so as to form a channel to catch any water which may beat in at the edges of the sash. This water discharges on the sill.

The bottom rail of the sash has a moulded drip let into it, so as to shed any water which may trickle down the outside surface of the sash. The under side of the bottom rail of the sash has an undercut; and directly under it, in the sill, a channel is cut. This channel catches any water which may beat in between the sash and the sill during driving rainstorms, and is discharged on the sill through perforations in the raised lip of the sill, as at **X**. These perforations consist of holes bored and reamed smooth. Three are usually provided for a window of ordinary width.

The frame and sash are constructed and set so as to form a reveal on the inside of the window; and when this is required, care should be exercised to allow ample space for window

shades between the inside surface of the sash,
when open, and the jamb lining. To secure this

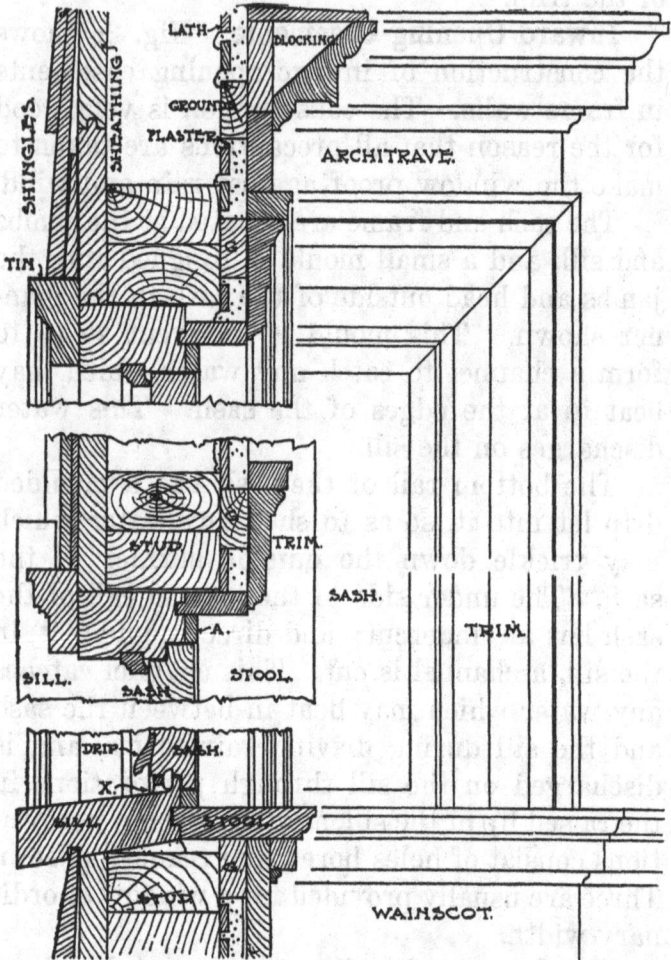

Fig. 26. Framing for Inward-Opening Casements.

space, it is usually necessary to thicken the jamb
of the frame and provide the filling piece **A**.

Shades for inward-opening casements are usually placed on the sashes, and the filling piece should be of a slightly greater width than the thickness of the shade when rolled up. The shade can, of course, be placed on the jamb of the window above the sashes; but this necessitates rolling the shade up entirely whenever the windows are to be opened.

Storm sash, blinds, or insect screens may be hung on the outside casing of the window; but, if required, the sill should be rebated for them. The outside casing is moulded, and mitered at angles, and at the head is flashed with tin, copper, or other suitable sheet metal carried up about six inches behind the shingles. The sill is grooved on the under side for the shingles, and on the inner edge for the inside stool.

The wall is constructed in the usual manner of studs, sheathed, papered, and shingled on the exterior, and lathed and plastered on the interior. Grounds (**G**) are set wherever necessary for a nailing for the trim, base, wainscot, and other interior finishing woodwork.

The trim is moulded and hollow-backed, and at the angles is put together with slip tongues or dowels, and mitered. A moulded back-band follows about same. It has an architrave head, and the fascia is plain and blocked at the top for the crown-moulding. This moulding may be cut out of one and one-eighth inch stock, or built up of several members. Crown-mouldings, except where they abut the ceiling, should be capped on top; otherwise they form a lodging place for dust and dirt.

The trim finishes on a stool with moulded edge. This stool, with its bed-moulds, corresponds to the cap of the wainscot, which is shown skirting the room. Walls behind wainscots, base, trim, and other interior finishing woodwork, should be plastered to the floor, but the white coat may be omitted. The jamb and head linings are tongued into the frame.

A simple inward-opening casement is illustrated in Figs. 27 and 28, adapted to the several types of wall construction, masonry, brick veneer, and stud frame.

Many architects think it necessary to rebate their casement sash. One carpenter who has employed this type of casement with all success for twelve years, has found in practice that rebating the sash simply weakens it and adds practically nothing to its tightness against weather. Rebated sash are objectionable because of the difficulty of fitting them accurately and the difficulty of refitting old sash which have, for any reason, ceased to fit properly through shrinkage, swelling, settlement, sag, or slight distortion of the sash.

Casement sash should be strongly made. For good work, sash should be 1¾ inches thick, and the sides and top 3 inches wide exclusive of glass rebate; the bottom members should be 1 inch wider, and all made of clear white pine

stock thoroughly tongued and pinned together.
There is little danger that sash so made will
ever sag. Large-sized butts should be used,

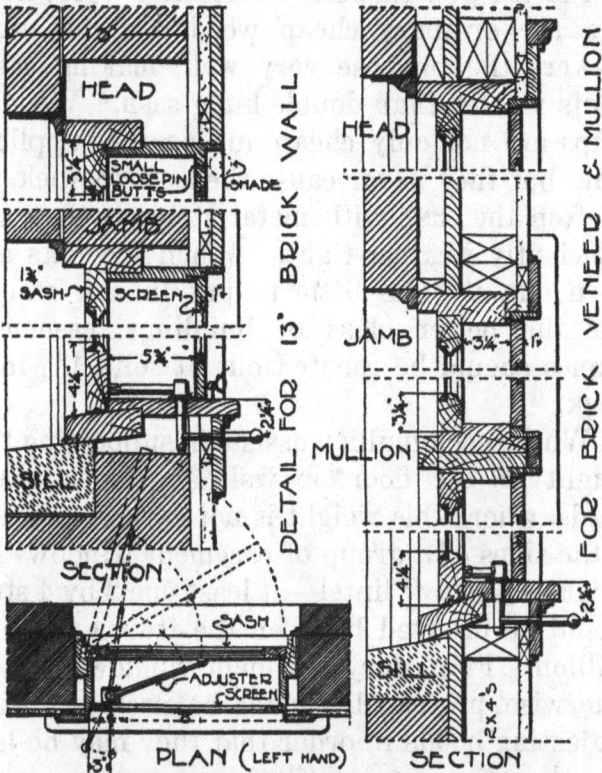

Fig. 27. Framing for Inward-Opening Casements.

preferably 4 by 4, galvanized, with brass pins
to insure their always working easily.

Except in the South and in California, casement windows should be equipped with weather
strips. If it can be afforded, some good form

of metal weather strip should be used. Both interlocking and friction strips are sufficiently tight.

For the general run of residence work, however the ordinary cheap wood and felt strip answers the purpose very well, making casements tighter than double-hung sash. Wooden strips are not only cheap and easy of application; but they never cause the sash to stick, as is often the case with metal strips if they are sufficiently weather-tight. When the felts are worn out, it costs little to put in new strips. For the better class of buildings, casement frames should be rebated out of solid $1\frac{3}{4}$-inch stock.

Where each mullion assists in supporting the weight of the floor or wall above—in other words, where this weight is not entirely carried to the sides of a group of casement windows by means of a heavy lintel—at least one 2 by 4 stud should be inserted between the frames of each mullion. First-story casement windows, unless otherwise protected, should be provided with projecting hoods in order that they may be left slightly open for ventilation in warm, rainy weather. The same is true of second-story casements in gables or elsewhere where not protected by the eaves.

The best method of fitting screens and storm sash to casements is to hinge them at the top, fitting them with a small bolt or catch to hold them in position at the bottom so that they will

not be blown in by the wind. Thus fitted, they
can be readily opened to give access to the hook
or adjuster by which the sash is held open, without interfering with curtains or shades.

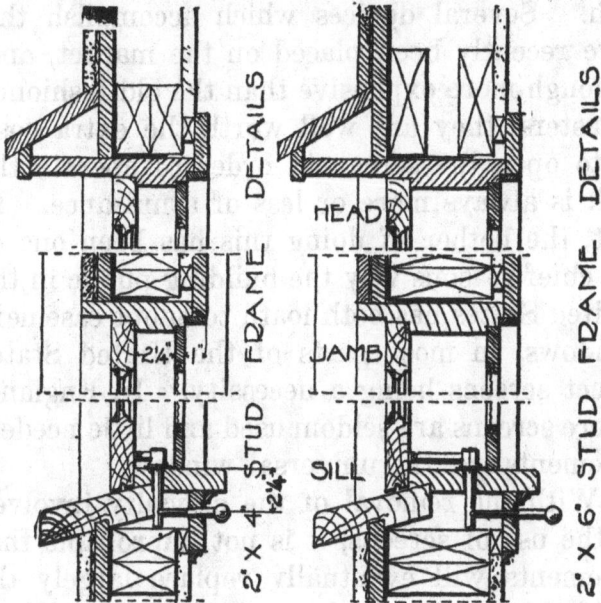

CASEMENT WINDOW DETAILS
HEADS JAMBS & SILLS 1¾" STOCK
SASH 1¾" THICK 3¾" WIDE OVER ALL
BOTTOM RAILS 4¼" " " " "
SCREENS ⅞" STOCK HUNG AT TOP IN
SMALL LOOSE PIN BRASS BUTTS

Fig. 28. Framing for Inward-Opening Casements.

The accompanying details all show sufficient
space in the jamb-lining next to the screen for
window-shade fittings, which are much neater
so applied than on the face of the casing. They
show the application of one of the new forms

of casement adjuster which operates the sash through a locking plate secured to the apron, doing away with the necessity of opening the screens in order to open, close, or adjust the sash. Several devices which accomplish this have recently been placed on the market, and, although more expensive than the old-fashioned adjusters, they are well worth the extra cost, as to open the screen in order to operate the sash is always more or less of a nuisance. In fact, the bother of doing this has been one of the chief reasons why the building public in the United States has both loath to adopt casement windows, in most parts of the United States insect screens being a necessity. In England, where screens are seldom used and little needed, casements are the universal window.

With the removal of the difficulty involved in the use of screens, it is not improbable that casements will eventually replace largely the old-fashioned double-hung sash for residence work, as most people appreciate their beauty as well as their great superiority as ventilators, particularly during warm weather.

Transom Window-Frame. Fig. 29 shows a detail of a transom window-frame, allowing for a three-inch bar with hinged transom light. This has the appearance of a moulded transom bar. The sash being hung the same as in the ordinary window, it allows it to be lowered at will, leaving no open joints for the wind and rain to get in when closed. It is well-nigh

impossible to construct a transom when hung
with butts, and have it weatherproof. In such
cases, the most satisfactory way is to hang the
sash at the bottom similar to that shown at the
right in Fig. 29. This makes a fairly tight job
and is simple in construction.

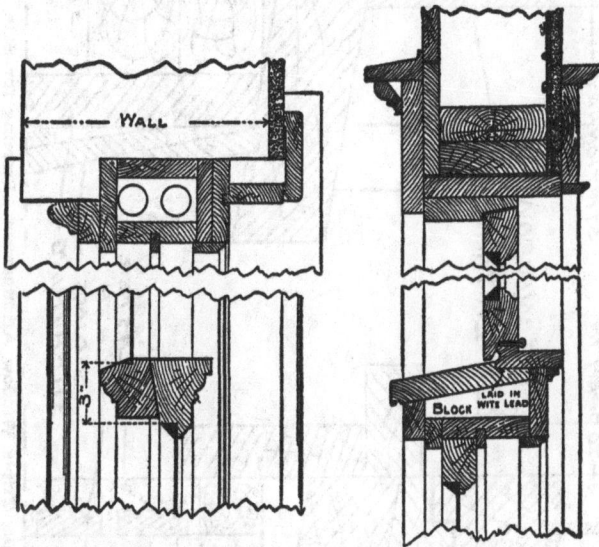

Fig. 29. Framing for Transom Windows.

A Triple Window. A triple window, with
two side sash casements, while the center sash
is stationary, presents an interesting problem
in window framing.

Fig. 30 shows detailed sections through the
sills of both the center and side windows. Both
the outside sill and the inside stool continue
through in line, and should be in one piece;

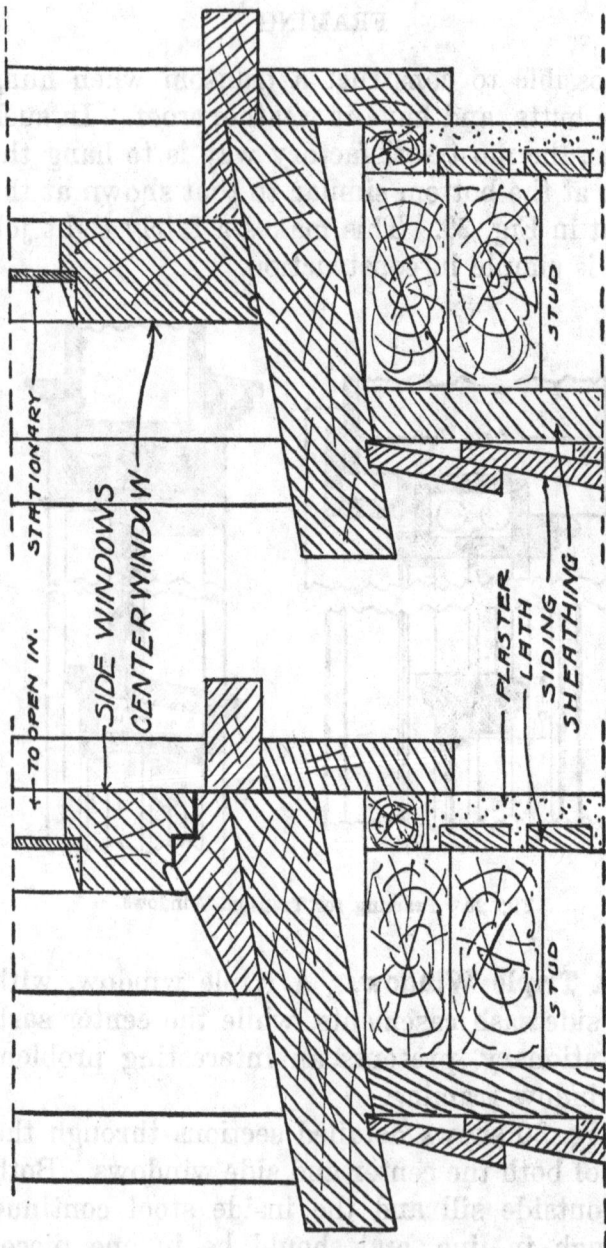

Fig. 30. Triple Window Arrangement—Framing for Side Casements and Center Sash.

STATIONARY

TO OPEN IN.

SIDE WINDOWS
CENTER WINDOW

STUD

PLASTER
LATH
SIDING
SHEATHING

STUD

and the casements have an additional sill, as
shown, which raises the sash sufficiently to clear
the stool. The writer has used this method
many times in actual practice, and same has
always given satisfaction.

Framing for Octagon Bay. Another point
in the construction of windows is in **bay windows** of the octagon pattern. Many bay win-

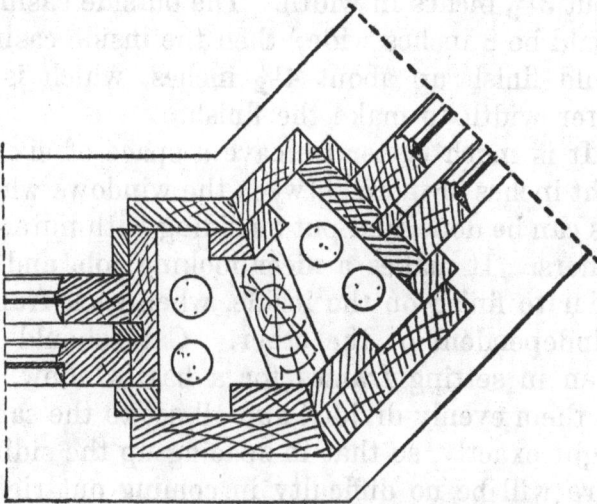

Fig. 31. Construction at Corner, Octagon Bay.

dows are made so that the casings join in the
angle. When casings join in this manner, it
is necessary to know just how to set the studding, and just the exact width of outside casing
necessary to use to make room for the weights,
and have it finish up right on the inside.

Referring to Fig. 31, it will be seen that it
requires an outside casing at least 7 inches wide

to get the frame in; and then the studding must be set directly in the center of the angle. Wherever the frames join as in this case, it is best to set a 2 by 4 stud in the angle as shown. It prevents the frames from drawing apart in the miter. The sketch shows that 7 inches is the least width of casing that will do; and this would leave the inside casing to finish up only about 3½ inches in width. The outside casings should be 8 inches wide; then the inside casing would finish up about 4½ inches, which is a better width to make the finish.

It is much better to leave a space of six to eight inches or more between the windows when this can be done, and put on siding with mitered corners. It makes a nicer looking job, and is easier to finish on the inside, when each frame is independent of the other. Care should be taken in setting frames for a bay window, to get them evenly divided and all set to the same height exactly, so that in spacing up the siding there will be no difficulty in coming out right.

A Simple Window Ventilator. Among the several theories as to the proper method of ventilating rooms where numbers of persons may be congregated at one time, there is one detail upon which the advocates of the several systems are agreed—namely, that the supply of fresh air should be admitted above the heads of the occupants of the room. This applies equally to warm-air or cold-air inlets, the reason being that the fresh air rapidly mixes with the air in

the room without causing a draught to strike
the occupants.

There are many houses, schools, and other
premises where regular systems of ventilation
will never be installed, but yet which should be
supplied with some means of changing the air
in their rooms. In the course of his professional

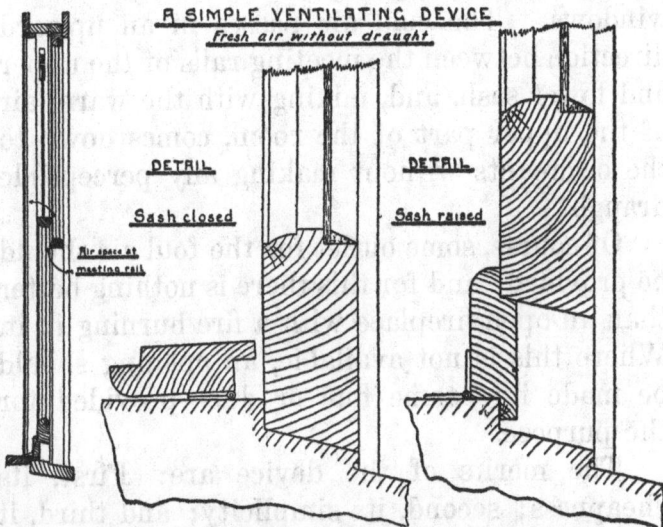

Fig. 32. Arrangement for Window Ventilating.

duties, the writer has been able to meet the
needs of many rural and small town schools for
some cheap and easy means of ventilation, by a
simple device which is illustrated in Fig. 32.

It has, of course, nothing new in the idea,
for most writers on ventilation have suggested
that a strip of wood could be employed at the
bottom of the window-sash, so as to admit air

at the meeting rails. The writer has found it best, however, to hinge the strip to the stop or to the window-board, and also to rabbet it as shown.

When hung in this way, it is always ready at hand, and cannot be taken out and lost. The rabbets are useful in preventing any draught from blowing in upon persons sitting near the windows. The cold air passes in an upward direction between the meeting rails of the upper and lower sash, and, mixing with the warm air of the upper part of the room, comes down to the occupants without making any perceptible draught.

Of course, some outlet for the foul air should be provided; and for this there is nothing better than an open fireplace with a fire burning in it. Where this is not available, an opening should be made into some flue or duct provided for the purpose.

The merits of the device are: First, its cheapness; second, its simplicity; and third, it is always in sight, and therefore its operation cannot be tampered with by ignorant persons.

Window Framing for "Open-Air" Room. How an apartment may be constructed and arranged to give the fresh-air advantages but none of the hardships of the ordinary outdoor sleeping place, is an interesting problem. For most of us—working more or less out of doors— the day-time supply of fresh air is all that it ought to be. It is with the air we breathe at

night that there is often room for improvement.
This point has been most emphatically pre-
sented during the past three or four years by
various eminent physicians, and by societies for
the prevention and cure of tuberculosis. It has
been proved conclusively that the germs cannot
develop or live in the presence of fresh, out-of-

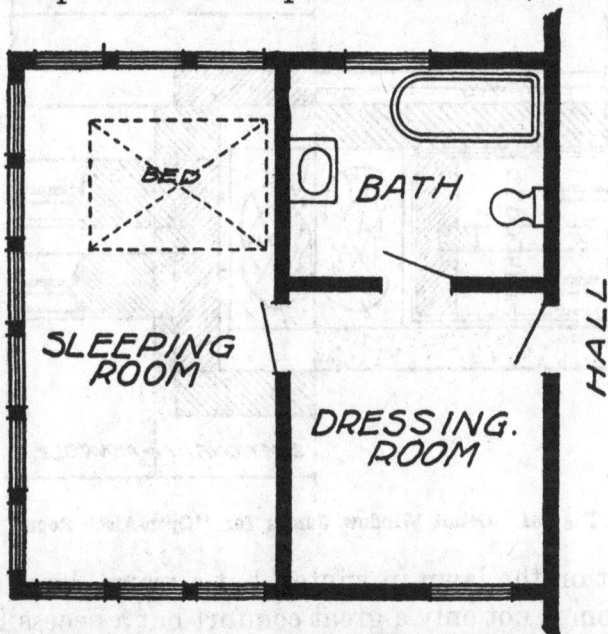

Fig. 33. Plan, Second Floor—"Open-Air" and Adjoining Rooms.

door air; but they do thrive, multiply, and
flourish indoors, in unventilated rooms. Stuffy
sleeping rooms are their especial romping place.

Fig. 33 shows the plan of a well-designed
"open-air" sleeping room. It is on the second
floor, open three sides to the weather. The
openings, however, are fitted with sash glazed

with tinted ondoyant glass to close when the
weather is extremely windy or stormy. Next
to the sleeping room, is a dressing room and
bathroom, both of which are kept comfortably
warm; but the sleeping room has no artificial
heat. A person can sleep comfortably in a tent

BELOW SILL＊ABOVE SILL

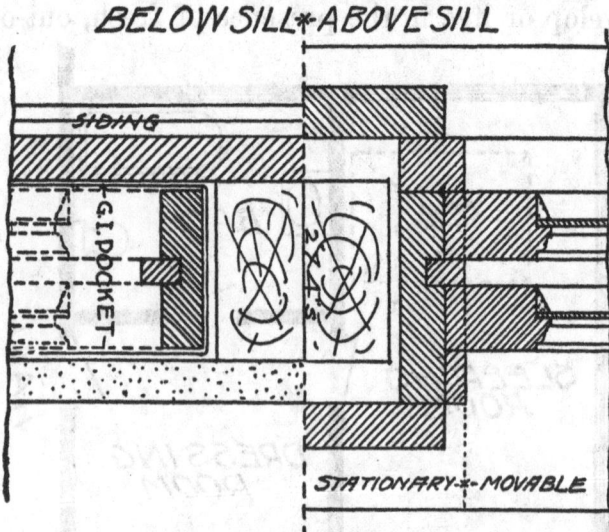

Fig. 34. Detail Window Jambs for "Open-Air" Room.

out on the lawn in winter, but a warm dressing
room is not only a great comfort but a necessity.

All the windows have curtains made out of
cotton duck, which may be drawn down and
buttoned to stud buttons screwed into the
casing. In summer the windows are all left
open, the only protection being the outside wire
fly-screens. On windy nights some of the canvas curtains are lowered and fastened to prevent
the wind from blowing in too strongly.

With rooms below on the first floor, great care has to be taken to guard against injury from sudden storms. The windows and floor of the sleeping room are very carefully made for this reason; in fact, you could play a hose on the floor of this room without in any way injuring the room below.

The detail drawings, Figs. 34 and 35, show how the windows are constructed to be perfectly storm-proof and water-tight. The jambs extend down, so that the sash may be lowered into pockets below the floor. These pockets are lined with galvanized iron, made water-tight, and

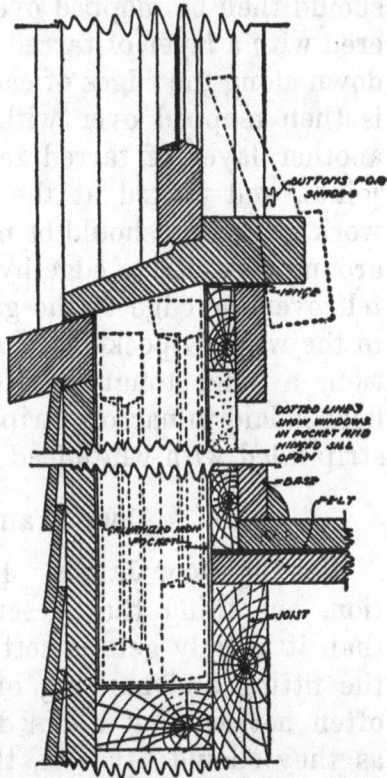

Fig. 35. Storm-Proof Box and Sill.

connected with a gutter outlet. All sash are balanced with heavy coil steel sash springs, so that the sash may be easily raised or lowered. When down, the opening is covered with a hinged cap.

In making the floor for such a room, great
care must be taken to have it water-tight. In
the first place, a matched pine floor should be
nailed to the joists in the usual way. This
should then be mopped over with tar, and covered with a layer of tarred felt carefully tacked
down along the edges of each sheet. This layer
is then mopped over with fresh hot tar, and
another layer of tarred felt laid to break the
joints, and tacked at the edges. In all this
work, provision should be made for a good joint
around the outside edge by carrying the tarred
felt over the edge of the galvanized-iron lining
in the window pockets. On top of this foundation, a solid tongued-and-grooved white oak
floor is laid in narrow strips, the groove of each
strip filled with white lead before driving it up.

Door Framing

Setting Door Jambs. In building construction, one thing that deserves more attention
than it usually gets is setting door jambs, and
the fitting and hanging of doors. You have
often noticed that doors do not shut securely
as they should; that is, they stand in or out
at the top or bottom of the door frame, as if the
door was **in a wind,** or **in a twist.** Of course,
carpenters have a handy way of getting out
of this trouble by saying the door is in a twist,
and that they cannot help it. Sometimes this
may be the case; but more often the trouble
is in setting the door jambs.

To begin with, good, straight studding
should be selected to set next to the door jambs;
but how often we find that the poorest studding
brought to the job is used in the partitions next
to the door. This is owing to the fact that the
partitions are the last place where studding is
used; and by the time the partitions in the house
are to be set, all the good studding has been
picked out and used, and nothing is left but the
crooked studs.

Scarcely anyone cares to use crooked stud-
ing; so he always picks out the straight studs,
until at last nothing but the crooked ones
remain. This may be used without harm in
many places about the building—for example,
for short rafters, headers, over and under windows, and for lookouts in the cornice. The contractor should see that they are used in these
places, and enough good, straight studding
reserved to use around the door openings next
to the door jambs.

The trouble with doors being in a twist more
often comes from the door jambs being set out
of plumb. If crooked studs are used, it is difficult to plumb them up securely, and two-thirds
of the carpenters will get the studding more or
less in a twist, or, as it is commonly called, **in
a wind**. When the man who sets the jambs
comes along, he will nearly always set his jambs
to conform to the plastering; otherwise the
jamb would stick out too far on one side of the
door, and not far enough on the other. It is

the edge of the jamb that is mostly neglected
and gives the most trouble. Many workmen
set the jambs with the wall, regardless of
whether they are plumb or not.

In Fig. 36, at **E**, is shown a jamb set with
the edge of one jamb out of plumb. Now, this
is just what is done about
six times out of seven,
where the doors seem to
be in wind. If the jambs
were set as shown in the
sketch, the door would not
close well at the top, but
would stand as shown at
T. The dotted line shows
how much this line is out
of plumb. In case it is
found impossible to get
the jambs exactly plumb,
do not set the jamb plumb
on one side of the door,
and the other side out of
plumb, as shown in the
sketch. Be sure that you
have both jambs alike.
If one is out of plumb a

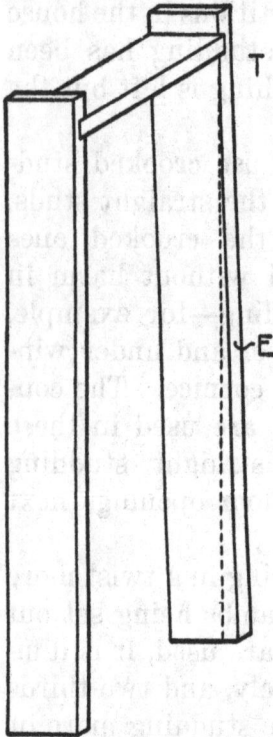

Fig. 36.   Door-Frame Jamb
        Out of Plumb.

trifle, be sure the other
one is out t h e s a m e
amount and in the same direction. Then your
door will not come in wind, and if the amount
is small no one will ever notice it.

Door jambs properly set facilitate hanging

doors, and a man can fit and hang more doors in a day to jambs that are set right. Most doors run one-eighth of an inch wider than the listed size; thus a door listed 2 feet 6 inches wide will in most instances measure 2 feet $6\frac{1}{8}$ inches; and it is safe, therefore, to make the jambs one-eighth of an inch wider than the width of the door. It makes a difference in the time of fitting the door when there is considerable to plane off, or if the door is so much wider that it has to be ripped.

Framing for Exterior Door. Fig. 37 shows the construction of an exterior door and frame in a frame house. The studs are doubled about the opening; and grounds (**G**) are set on same for the trim and the door jamb and head. The jamb is of seven-eighths-inch stuff with a moulded stop planted on. Another way of constructing the jamb is to get it out of one-and-one-eighth-inch stuff and rebate it for the door, thus doing away with the applied stop.

The door is veneered on both sides one-eighth inch thick, and on edges five-eighths inch thick, on a core of white pine strips glued together with the grains reversed. The core is frequently tongued and grooved together. Thicker edge veneer is required so as to permit of planing down the edges without exposing the core. In the better class of work, the face and edge veneers are mitered at the angles so as to conceal the end grain of the face veneer.

Panels should be set loose so that expansion

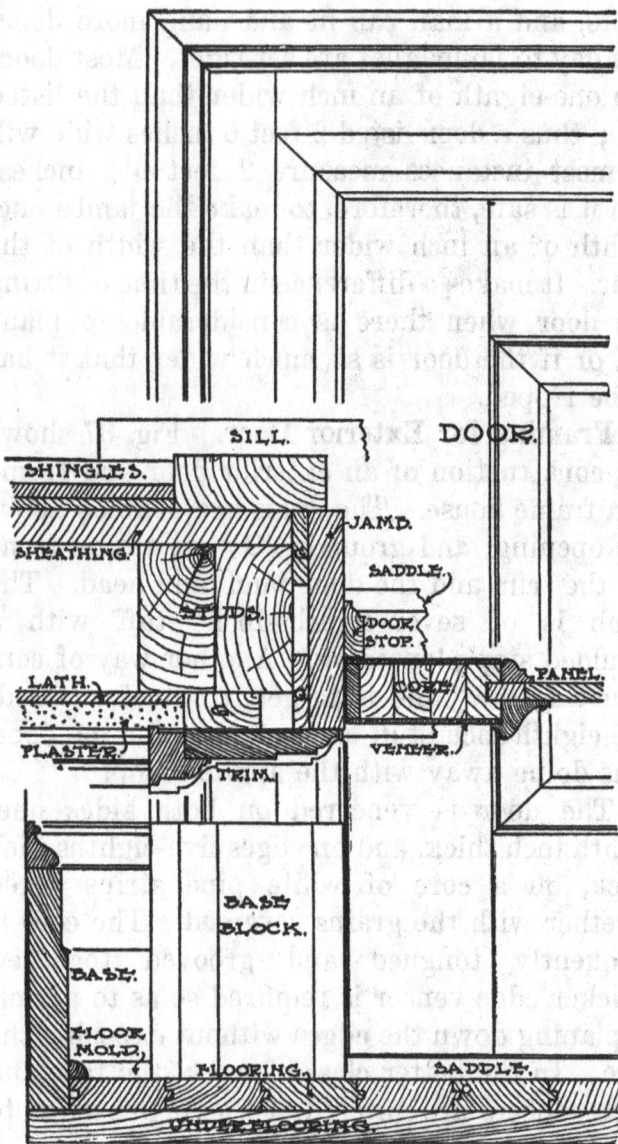

Fig. 37. Framing for Exterior Door.

and contraction will not split them. The mouldings are nailed to the stiles and rails and a pine fillet is set in between them.

The trim is worked out of seven-eighths-inch material, and has a moulded back-band. The little tongue left on back edge of back-band may be planed to fit the unevenness of the plaster.

Hanging Sliding Parlor Doors. There are now many patented hangers for sliding doors on the market, each possessing more or less

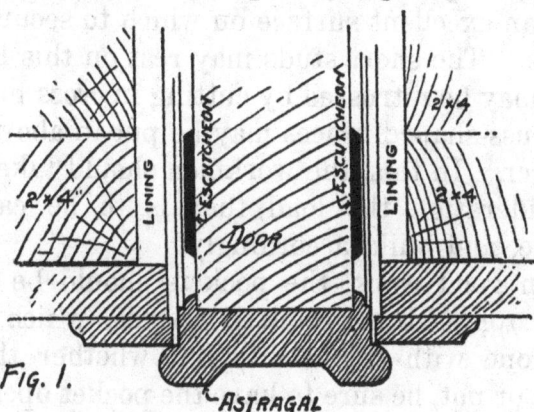

Fig. 38. Sliding Doors—Construction at Jamb.

merit. Full directions are furnished with each set. Any average workman should be able to put up the work. The main thing is to see that the partition rests on substantial bearings to prevent settlement, as this will necessarily throw the track out of level and affect the free working of the doors. Be sure to set the studding plumb and properly spaced for the pocket. Never set the studding flatwise with the door.

Never allow a hot-air pipe to run up beside the sliding door, when it is possible to place it in some other partition. Always double the studding at the jambs; and be sure to make proper calculations for the opening so that when the finished work is in place, the full face of the door will show when closed. Be sure to have the woodwork over the opening perfectly rigid. Two well-seasoned joists spiked together and set up edgewise, make a good truss or lintel, and an excellent surface on which to secure the track. The short studs may rest on this lintel, and may be retrussed by cutting in cross-braces, or truss-shaped braces may be put in above the hanger. In this, the workmen should take into consideration the load that is to be carried above, and build accordingly.

In good work, the pockets should be lined with tongued-and-grooved boards, which may be done with thin stuff; but whether this is done or not, be sure to have the pocket openings cut off at the back end so that there will be no connection with other openings in adjoining partitions and outer walls. This should be done for several reasons—first, to heat the house, because these openings will create a draught; then again, if a fire gets started in a partition, these openings furnish an excellent draught to fan the flames.

Another point we might call attention to is the unsightly notching-out of the stops to allow the raised escutcheon to pass into the pocket.

This can be avoided by running a stop around
both sides of the door, and membering with the
astragal as shown in Fig. 38. The stops on the
jambs are set as shown. Thus, it will be seen
that the escutcheon is cleared; and that, when
the door is shoved back, the astragal will cover
the pocket opening, and to all appearance is
simply a mould made fast to the jambs. The

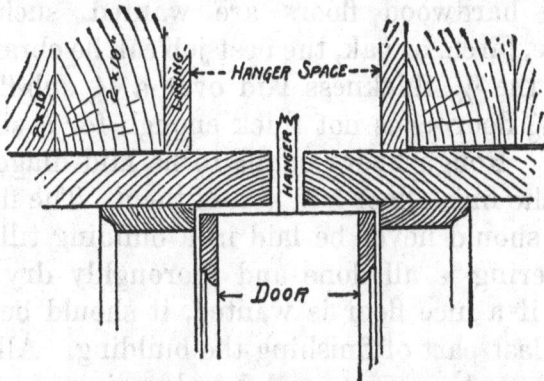

Fig. 39. Sliding Doors—Construction at Head.

head jambs should be set to allow only for the
free working of the hanger, as shown in Fig. 39.

Value of Sub-Floors

As far as warmth is concerned, there may
not be any necessity for the double floor for
houses in warm countries. There are other reasons for double floors besides the heat and cold
question. The rough floor is a great convenience to work on during the construction of
the building; and in case of a brick, stone, or
hollow block building, the rough floor is almost

indispensable. Even if one did do without it, there would be the labor of covering the greater part of the floor space with a temporary floor to work on during the construction of the building.

Then, again, the finish floor cannot be laid in advance, so that it can be used to work on; for if so, it would be spoiled, and unfit to be seen when the building was completed.

If hardwood floors are wanted, such as maple, birch, or oak, the best job can be obtained with the $\frac{3}{8}$ thickness laid over a $\frac{7}{8}$ sub-floor; the $\frac{3}{8}$ flooring is not thick enough for a single floor. Rough floors ought to be laid diagonal, and the finish floor laid across them. The finish floor should never be laid in a building till the plastering is all done and thoroughly dry; in fact, if a nice floor is wanted, it should be the very last part of finishing the building. All the base-boards, casings, and door hanging—in fact, everything should be done before the finish floor is put down. Carpets are not used as much now as formerly, and nice floors are much desired; and these are not possible without the first rough floor to take the wear and tear of the building construction.

If anything like a good building is desired, it is false economy to dispense with the rough floor, no matter what the climate may be; hot or cold, the rough floor has merits that make it worth while to put in. It also adds largely to the strength of a building, both in the floor and to withstand wind pressure when it is a second-story rough floor.

How to Build a Screened Porch

Details for the enclosure of porches are given in Fig. 40. The question is sometimes asked whether it is best to use a screen made wide enough to cover the height of the porch in one piece, or to have the screens in sections.

The widest screen that is carried on the market is 48 inches; but so wide a screen could not be recommended, as it should be supported

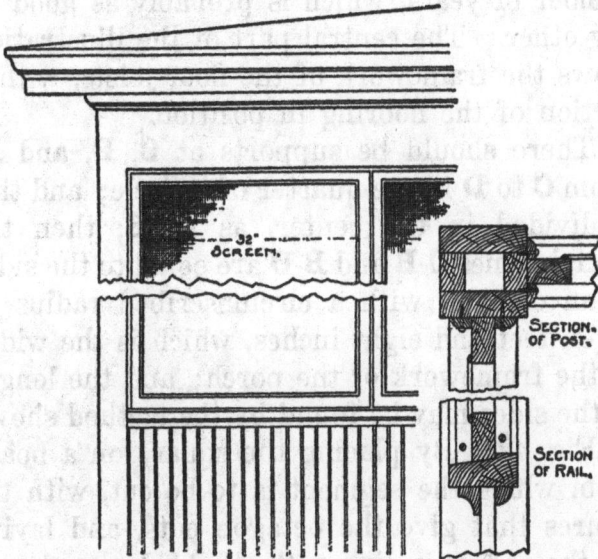

Fig. 40.   Construction and Design of a Screened Porch.

with cross-pieces, anyway. We recommend using frames about the size or width required for an ordinary window. Regular stops should

be put on with round-headed screws. The
frames can be easily removed in the winter-time;
and solid panels made of ceiling can be set in
their place, if desired. A few panes of glass set
in these panels will give sufficient light, and
thus make a very efficient storm-door enclosure.
The accompanying sketch shows about as good
a way as any for the construction of a porch
of this kind.

Constructing a Circular Porch

Fig. 41 shows a method largely used for a
number of years, which is probably as good as
any other. The central part of the illustration
shows the framework of the floor-joists, with a
portion of the flooring in position.

There should be supports at **C**, **B**, and **D**.
From **C** to **D** is one-quarter of a circle; and this
is divided in the center, as at **B**; then the
straight lines **C-B** and **B-D** are equal to the sides
of an octagon with a circumscribed radius of
seven feet and eight inches, which is the width
of the framework of the porch; and the length
of the sides may be found by the method shown
in Fig. 42. By placing the square on a board
from which the segment is to be cut, with the
figures that give the octagon cuts, and laying
off the radius in line with the blade, as shown,
describe the arc, and it is ready to cut. The
figures shown on the square will give all the
cuts required in the framework about the octagon, as the blade will give all the cuts at **B**,

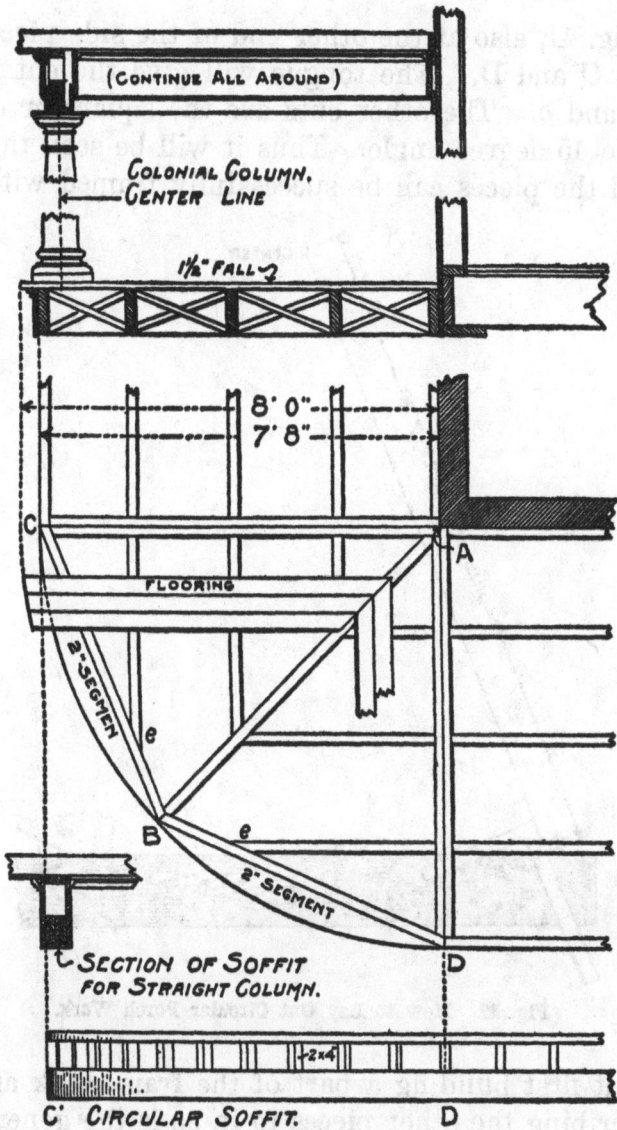

(CONTINUE ALL AROUND)

COLONIAL COLUMN.
CENTER LINE

1½" FALL

8' 0"
7' 8"

C

A

FLOORING

2" SEGMENT

e

B

e

2" SEGMENT

SECTION OF SOFFIT
FOR STRAIGHT COLUMN.

D

2×4

C' CIRCULAR SOFFIT. D

Fig. 41. Framing for Circular Porch.

77

Fig. 41, also at the other end of the side pieces
at **C** and **D**. The tongue will give the cut at
e and **e**. The other cuts are the square or on
the 45-degree angle. Thus it will be seen that
all the pieces can be successfully framed with-

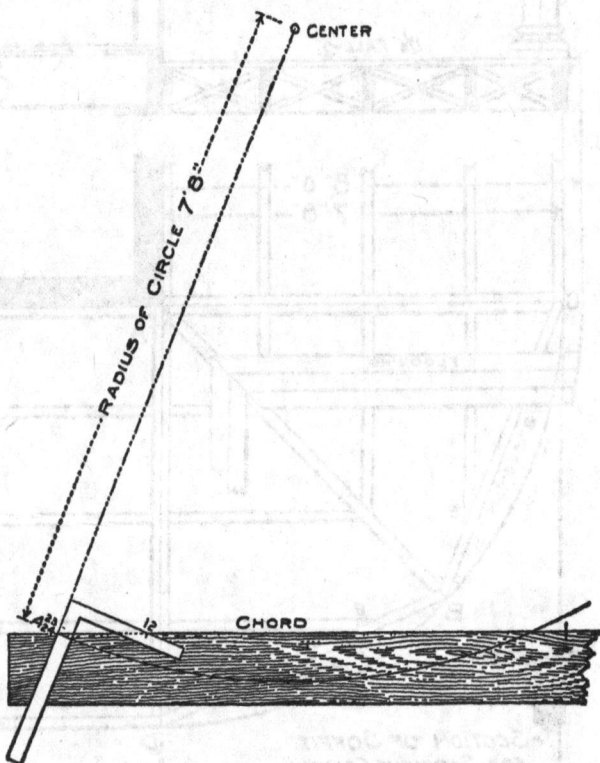

Fig. 42. How to Lay Out Circular Porch Work.

out first building a part of the framework and
scribing the other pieces to it, as is the general
custom.

There should be four of the segment pieces

got out, setting one flush with the top edge, and one at the lower edge of the joists. The upper ones should be of one and three-fourth inch stuff, same as the joists, while seven-eighths will be sufficient for the lower member. Set blocks between these segments, nailing them well to the joists; also set a few blocks flush with the face of the segments, which makes an excellent form to secure the base.

The ceiling joists are usually put on the narrow way of the porch with an angle piece same as at **A-B**, on which to form the miter joint of the ceiling.

To form the soffit, use seven-eighths by six or eight inch sized boards, and spring them to their proper place just the same as building a circular girder. The first board should be sprung to a form, and the next board well nailed to this one, and so on, till the soffit is the required thickness or strength. It is not always necessary to build to the full width desired, as it can easily be furred out to the required width. The soffit should be continuous—that is, for the straight part, as well as for the circle. Long boards should be used so as to lap well around the circular part, being careful not to break joints on the circular part or at **C** or **D**.

A soffit, if properly built in this way, will not necessarily need a column set at **B**, as it will be self-supporting. If straight columns are used, the outer face of the framework should be flush with the framework below; but if tapered or colonial columns are to be used, then the center of the soffit should rest over the center of the column, as shown in Fig. 41.

In case a deep frieze is wanted, it may be had by building on top of the soffit girder with

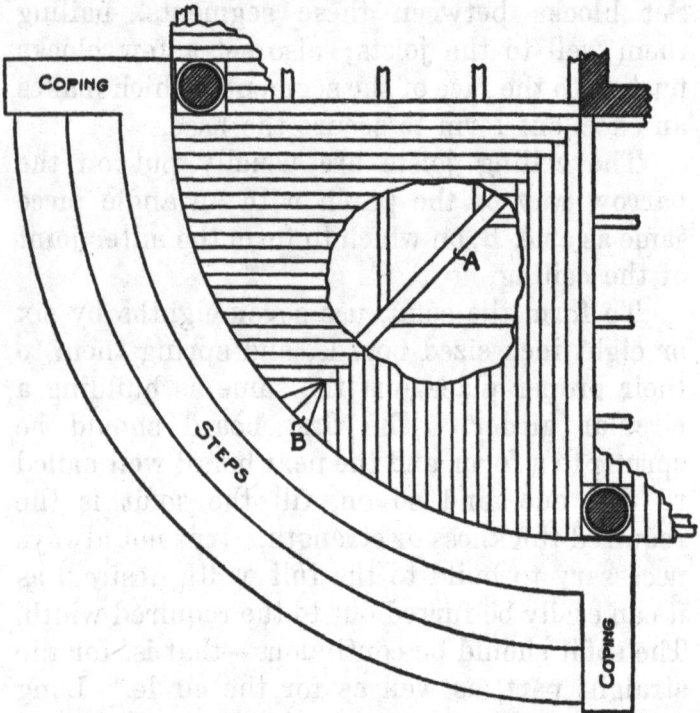

Fig. 43. Floor Construction—Circular Porch.

blocks, and putting a formed plate on these. For all circular mouldings, it is better to have them solid, and they will then always stay in place, as there will be no kerf joints to open up after the work is completed.

Another circular porch is shown in Figs. 43

and 44. It looks better than the mitered seam, is perhaps as good, and is cheaper.

At **A**, Fig. 43, is a 4 by 6-inch timber, which gives a good bearing for the ends of the flooring boards. **B** shows the method of finishing the

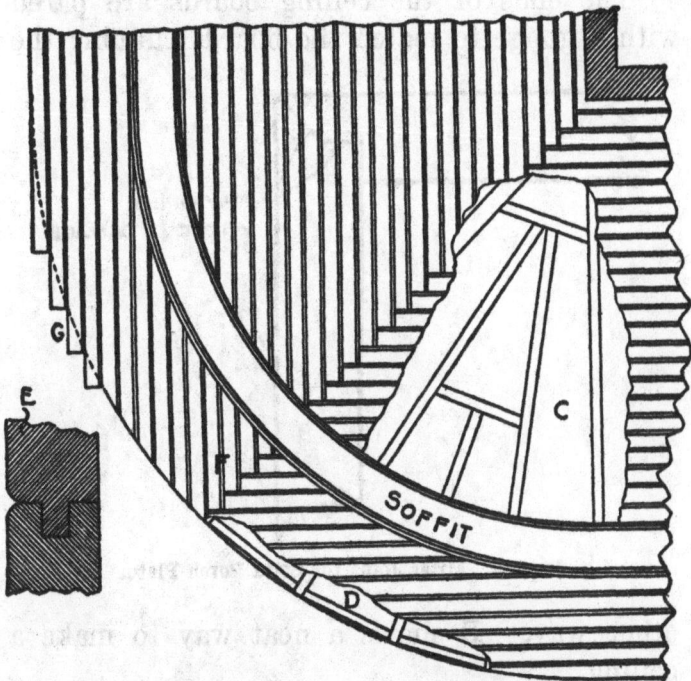

Fig. 44. Ceiling and Roof Framing—Circular Porch.

floor where steps come on the circle.

Opening **C**, in Fig. 44, shows how the heels of the ceiling joists are put in to give support for the heels of the rafters.

F in Fig. 44 shows the method used for put-

ting a ceiling plancher on a quarter-circle porch.
D shows cripples fitted in between the joists to
which the ends of the ceiling boards are nailed.
Put the plancher on, and saw to the circle afterwards as shown at **G**.

The ends of the ceiling boards are pared
with a gouge to match the boards running the

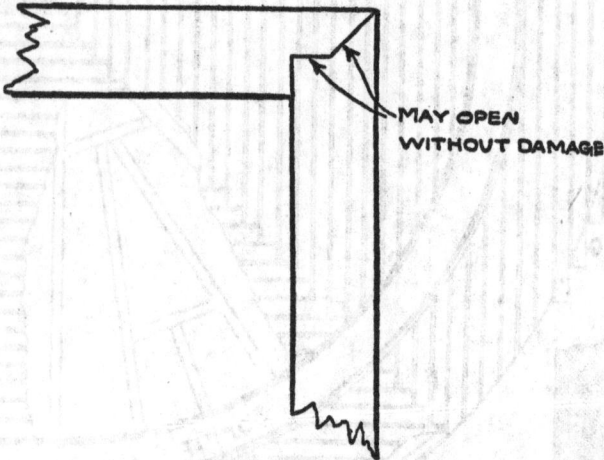

MAY OPEN
WITHOUT DAMAGE

Fig. 45.　Miter Joint for Solid Porch Plate.

other way.　**E** shows a neat way to make a
ceiling.

A good **miter joint** for a porch plate is shown
in Fig. 45.　It is especially good for the miter in
a solid porch plate, or other places where it is
desired that the mitered sides shall show and
where the edge cut is not exposed.　The timbers
are often crooked, and this joint will allow for
considerable variation from the square.

Framing for a Fireplace

In Fig. 46, two flues are shown, one to extend to the basement floor, and they are for use of stoves in adjoining rooms. When thimbles are put in to make connection with adjoining rooms, the brickwork should be corbeled out to the full thickness of the wood partition, and a long thimble used to extend through the brickwork, being careful not to let the thimble protrude into the flue space. At sketch **A,** another way of widening the brickwork at the thimble is shown, which is simply to cut in a cross-piece between the studding, and on this build the extra brickwork with all joints well filled with mortar. In all cases the thimbles should be set at the time of building the chimney, being careful that all joints are well filled and tuck-pointed on both sides; and in addition to this, it would be well to plaster on the inside of the flue from bottom to top.

In the illustration we have shown an ash-pit beneath the fireplace, where the ashes may be dumped and taken out later. This pit should have a vent into the flue, so that when the ash dump is opened a downward draught will be created which will prevent the ash dust from flying back into the room. For supporting the hearth, use iron bars made of $\frac{1}{2}$ by 2 iron; and on this, lay brick edgewise, leaving a space of three or four inches for concrete on which to lay the tile hearth. The fireplace should be lined with firebrick, with the upper part of the brick slanted toward the top of the opening, as shown in the cross-section. The arch in front should be supported on a segment made of ⅜ by 3 inch wrought iron, set back from the front so that

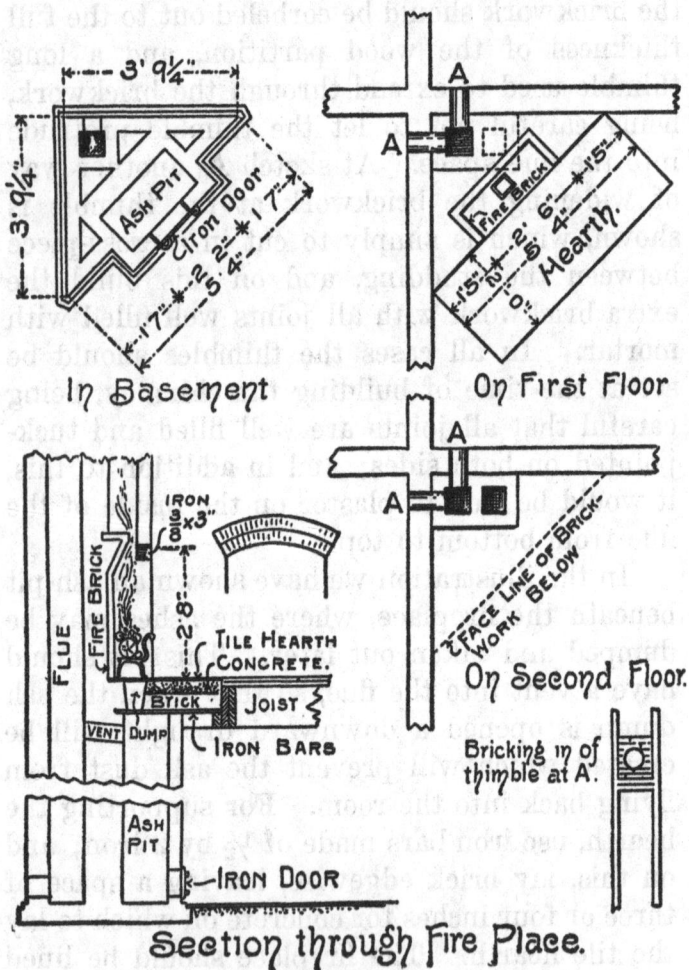

In Basement

On First Floor

On Second Floor.

Bricking in of thimble at A.

Section through Fire Place.

Fig. 46. Arrangement for Fireplace.

it will not show. If a straight-top opening is
desired, then use a 3-inch by 3-inch angle iron,
with the flange on the inside of the brickwork.

The dotted lines show the position of the
flue for the fireplace, and will require the opening or throat to draw over to it; but it should
start straight from the fireplace, and gradually
draw over to its position as shown. The face
of the brickwork should carry up to the ceiling
of the first story; and this gives ample space
to make the proper bend in the flue. The flues
should be independent of other openings.

Cast-iron hoods with damper attachment are
quite often used to form the top of open fire-
places, and are set in place at the time of building the chimney. The top should be capped
with Portland cement, or with a 3 or 4-inch flat
stone with openings cut to fit the flue openings.

Constructing a Cupboard

Fig. 47 is a sketch of a cupboard. It makes
a very complete fixture for the kitchen, and is
easily made. The flour bin is hung the same as
a door, and swings outward as shown by the
dotted lines in the plan. It is quarter-circle in
shape, the back being made of zinc, with a roll
rim at upper edge. A board $1\frac{1}{8}$ inches thick
and 16 or 18 inches wide is used for the front
and back. The bottom is grooved in about $1\frac{1}{2}$
inches from the bottom ends of these pieces,

GLASS. GLASS

PANEL PANEL

BIN DOOR DOOR

CUPBOARD.

ZINC WITH ROLL RIM.

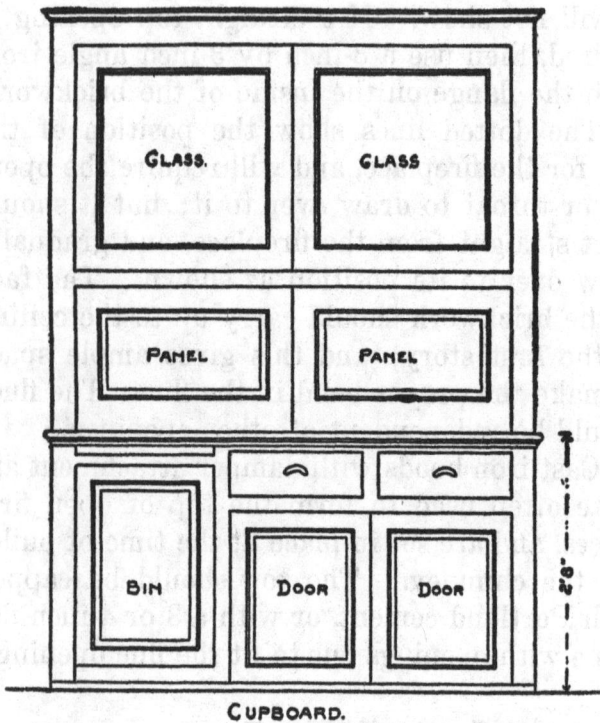

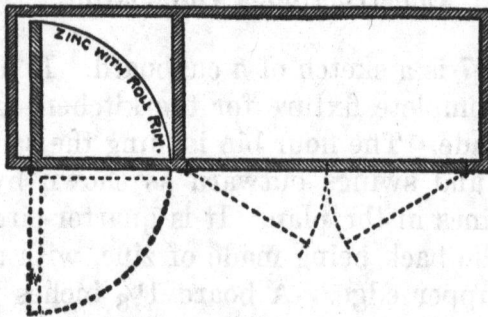

Fig. 47. Design for Kitchen Cupboard.

and gained in about ⅝ inch deep. The front
and back edges are grooved about ⅝ inch wide

by ¼ inch deep, to receive the zinc back, which is made fast by nailing with round-headed tacks. The front may be paneled, as shown in the drawing, if a better finish is desired.

Shingling the Sides of a Building

It is the style in many sections of the country, to shingle the sides of buildings, not only for the smaller class of buildings, but for public buildings as well. See Figs. 48, 49, and 50.

When properly done, it makes a good-looking building, and the cost is generally less than for any other siding material, since wall shingles can be put on with more space exposed than in regular roof work; also a cheaper grade of shingles can be used for this purpose, with good satisfaction. One thousand will cover about 150 square feet of surface; and a man will put on about as many in a day as he can on the ordinary roof in the same length of time. As this kind of work is comparatively new, differing in some respects from roof shingling, it may be that some will be benefited by a few pointers concerning the work.

The building may have corner boards and water-table, though these are generally omitted. It makes a more weatherproof and at the same time a better-looking job, to run the shingles out to the corners. The first course at the bottom should be double. The first or under course furnishes a good place to work in some of the poorer shingles. In shingling the corners, the shingles on one side should be kept flush with the corner, and those on the other side should be flush with the butts or a little beyond, so that they may be trimmed even by sawing in on the edge and cutting out with a knife.

There are several ways of getting the first course straight; a straight edge can be used by tacking a shingle at each end to hold it in place, and should be used for each course thereafter.

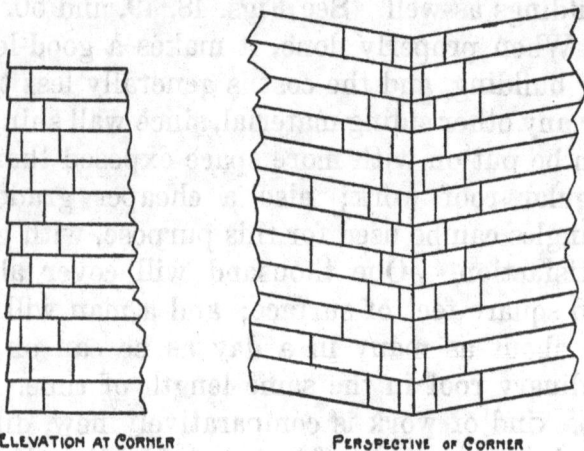

ELEVATION AT CORNER   PERSPECTIVE OF CORNER

Fig. 48. Shingling a Corner.

The siding boards should be straight and level, and the first course should extend a little below to form a water drip. The courses should come even with the top and bottom of the window-frames, which can be easily done by varying the courses the same as in clapboard siding. It will be necessary to cut the tops of the two last courses under the windows, but the pieces can be used at the tops of the windows, which should

have a rabbeted cap. The shingles should be
doubled at the point; one row should be put
above the cap, and the other should drop below
these. At the corners of the window-frames, it
is better to cut out a corner of a few shingles
so as to break joints.

When a first-class job is desired, it is better
to put the cornice on
and cut the shingles to
fit under it. For some
classes of buildings, it is
all right to put the cor-
nice on over the shin-
gles; or, in case the cornice is already on, to
drive the shingles up
under it.

Do not use too large
nails for shingling. If
the boards are sound,
three-penny nails are
large enough.

Paper may be put
under the shingles, and
in some cases this is the
best method; but as so
many nails are driven
through the paper, it seems better to put it on
the inside of the sheathing between the studs.

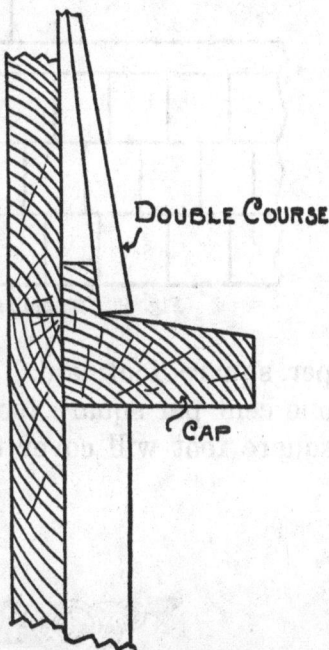

Fig. 49. Shingling Over a
Window.

As to the cost of this method of building,
boards for less than $20.00 per thousand may

be used, which will answer. The shingles at
$4.00 per thousand will come to about two cents

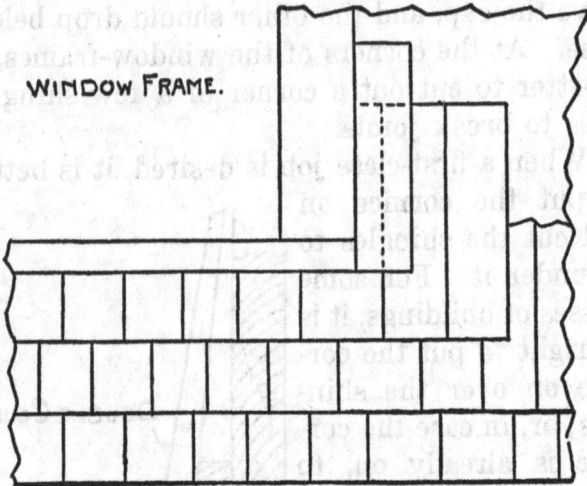

WINDOW FRAME.

Fig. 50.   Shingling Below a Window.

per square foot laid. Plaster will cost about
one cent per square foot, so that five cents per
square foot will cover the cost.

Roof Framing Simplified

The details of roof construction are comparatively simple, and are pretty generally understood. It is in the laying out of the different members, finding their proper lengths and cuts, that the difficulties of roof framing arise. Owing to the many different styles and pitches of roofs, this is considered a very complicated matter by a great many otherwise good mechanics, who accordingly resort to certain unsatisfactory "cut and try" methods and "rules-o'-thumb" to lay out the work.

The steel square is the carpenter's best assistant for laying out all framing; but it is for roof work that its use is most essential. By the use of the steel square—after a very few of the fundamental principles of roof framing are well understood—the whole subject becomes clear to an astonishing degree.

Roof Pitches and Degrees. Fig. 51 contains a whole volume on roof framing. The fractional pitch lines for the common rafter are shown for each inch in rise up to the full pitch; and their lengths are expressed in decimal figures to the one-hundredth part of an inch; while to the right of the blade, the same are expressed for the corresponding octagon and for the common hip or valley for a square-cornered building,

91

which are reckoned from 13 and 17 on the tongue respectively. However, neither is absolutely correct, though near enough so far as the

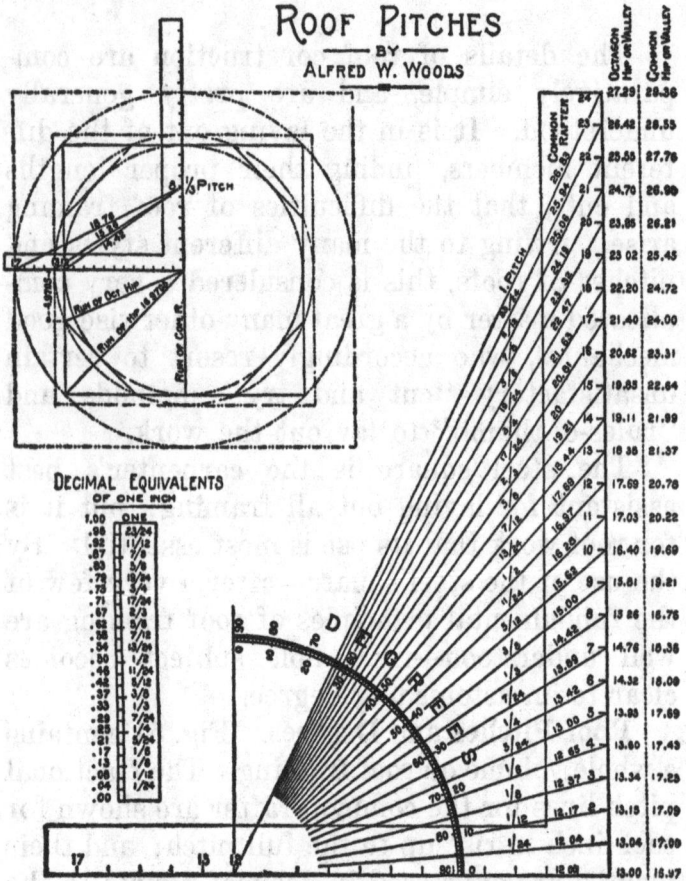

Fig. 51. Roof Pitches and Degrees on the Steel Square.

cuts are concerned, the greater deviation being in the hip for the square-cornered building. It lacks .0295 of being 17 inches, and represents the run of the hip to a 12-inch run of the common rafter. Its true length being 16.9705 inches, this is the length from which we have reckoned for the lengths of the hips, instead of 17, as is the usual custom. This may seem a trifling difference; and so it is in a short run and low pitches; but suppose it is for iron construction. To begin with, the shortage of each foot in run with the common rafter is .0295 inch; added to this is the gain it would have in the pitch, which would be .015 of an inch by the time it got up to the full pitch for the common rafter. This, added to the .0295 to start with, would be a difference of .0445 inch to the foot in run with the common rafter. Now, suppose the run to be 18 feet; 18 times .0445 equals .8 plus, or $^{19}/_{24}$ of an inch difference; or, if no account were made of the gain in pitch, the .0295 inch in the run would amount to over half an inch in the length of the hip alone.

This is a common error; and while it is not much, and probably would never be noticed in wood construction, it is well to know this discrepancy and guard against it when the occasion demands, and for that reason we give the correct amounts. The shortage in the octagon is not so pronounced. Instead of it being in the run, it is the tangent that is lacking the same amount, it being 4.9705 instead of 5 inches. This, coming as it does, cannot affect the length of the rafter nearly so much as in the above.

We explain this shortage better by referring

to that part of the illustration showing the plan
of a combination square and octagon frame with
the heel of the steel square resting at the center.
From this it will be seen that the two outer
circles catch the corners of the frame and seem-

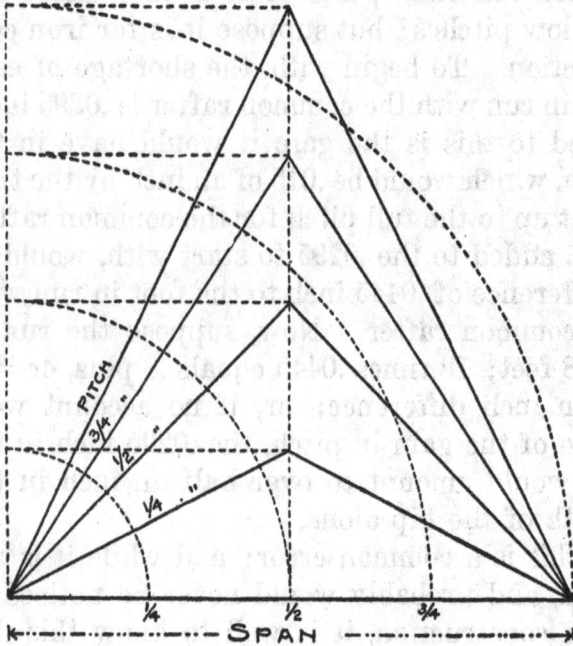

Fig. 52. Roof Pitches.

ingly intersect the tongue at 13 and 17, the
figures used on that member for the seat cuts;
but the true length of the run of the hip is
16.9705, and that for the tangent of the octagon
is 4.9705.

In connection with this illustration we also
give a table of decimal equivalents to the one-

twenty-fourth part of an inch, for convenience in finding their values in common fractions.

What Determines the Pitch? This is a simple question; yet there seems to be a wide difference of opinion as to what determines the incline given the roof. Custom has long since settled upon the rise given the roof in proportion to the span; thus, a one-fourth, one-third, one-half, etc., pitch, must have a rise in that proportion to the span. Reckoned on this basis, a full pitch has a rise equal to its span. See Fig. 52. Here the span is divided into several parts. The dotted lines are shown

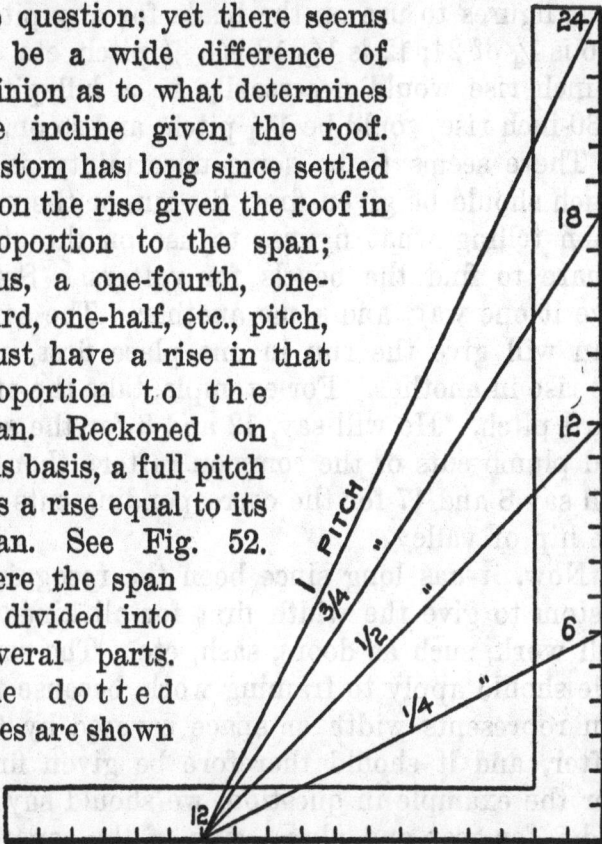

Fig. 53. Roof Pitches on the Steel Square.

in the transferring of these parts to the rise line. In Fig. 53 these parts are shown in connection with the steel square. Twelve is used on the tongue, because it represents a one-foot run.

The span would therefore be two feet, or 24 inches, which is equal to the length of the blade. It is then a very easy matter to fix in the mind what figures to use on the blade for any pitch, as 6 is $\frac{1}{4}$ of 24; 12 is $\frac{1}{2}$; 18 is a $\frac{3}{4}$ pitch, etc. A 24-inch rise would necessarily be a full pitch. A 30-inch rise would be $1\frac{1}{4}$ pitch; and so on.

There seems to be some uncertainty as to which should be given first, the run or the rise, when telling what figures to use on the steel square to find the bevels for rafters. Some give it one way, and some another. The same man will give the run in one place first, and the rise in another. For example, take the one-third pitch. He will say, 12 and 8 for the seat and plumb cuts of the common rafter; then he will say 8 and 17 for the corresponding cuts for the hip or valley.

Now, it has long since been the recognized custom to give the width first for all kinds of mill work, such as doors, sash, etc. The same rule should apply to framing work, because the run represents width or space covered by the rafter, and it should therefore be given first. For the example in question, we should say 12 and 8 for seat and plumb cuts of the common rafter, and 17 and 8 for the corresponding cuts for the hip or valley. It is better to always take the figures 12 and 17 on the tongue, because they are standard for any regular pitch; the blade will admit of from 1 to 24 inch rise per

foot, besides giving a greater range of side cuts
without change of figures on the tongue. Then,
again, it helps to familiarize the mind as to
which member of the square gives the desired
cuts.

How to find the cuts for rafters, no
matter what the pitch, is a point that
gives trouble, yet there is nothing simpler when properly understood. Take,
for instance, how to find the cuts for
hips, valleys, and jacks when the common rafters are 6 to 12, 7 to 12, 8 to 12,
9 to 12, 10 to 12, etc.

Take the first example, 6 to 12. The
formula given applies to all alike,
whether it be a six-inch or a fifteen-
inch rise to the foot. Fig. 54 will show
why certain figures are used on the
square to obtain the cuts. Of course
other figures can be used, but they must

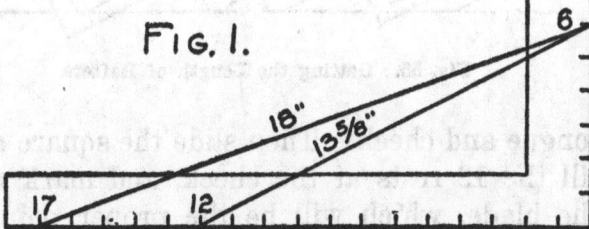

FIG. I.

Fig. 54. Lengths and Cut of Common and Hip Rafters.

be in the proportions here given. **Twelve on
the tongue is used because it represents one foot,
and 17 because it is the length of the diagonal**

of a foot square, and represents the corresponding run of the hip or valley to one foot run of the common rafter. These figures are standard or fixed points for any pitch desired.

Taking the 6-inch rise to the foot, the common rafter is 13⅝ inches, and the hip or valley 18 inches for a one-foot run. Now, suppose we wish to find the length of the common rafter for a building 22 feet 6 inches wide. Since the run is one-half of this amount (11 feet 3 inches), all that is necessary is to place the square at 12 and 6 along the edge of the rafter eleven times (see Fig. 55); and as there are 3 inches more, lay off that amount from 12 along the

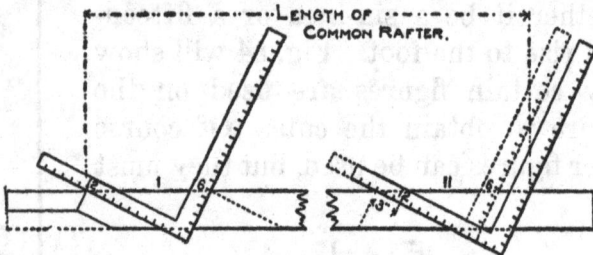

Fig. 55. Getting the Length of Rafters.

tongue and check. Then slide the square along till the 12 rests at the check, and mark along the blade, which will be the proper point for the plumb cut.

Proceed in like manner for the hip or valley, taking 17 and 6; but, at the last placing of the square, instead of measuring off 3 inches, take 4¼ inches, which is the length of the diagonal of a 3-inch square. This may be reckoned as follows: Since 3 inches is one-quarter of 12 inches, one-quarter of 17 inches equals $4\frac{1}{4}$ inches. Thus, the length of the rafters is obtained without any further measurement, and that, too, without knowing their actual length.

The jacks being a part of the common rafters, their lengths may be found in the same way. Or, if they are to set on 16-inch centers, place the square at 12 and 6, as for the common rafter, and mark along the tongue; then slide the square along till 16 rests at the edge of the rafter, and the length will be indicated by that part of the rafter covered by the square, which represents the common difference of the jacks.

However, if one is good in mathematics, it is often better to find the rafter lengths by multiplying the lengths for one foot by the run. Taking the above case: $11\frac{1}{4}$ times $13\frac{5}{8}$ inches equals 12 feet $9\frac{1}{4}$ inches, the length of the common rafter; $1\frac{1}{3}$ times $13\frac{5}{8}$ inches equals 1 foot 6 inches, the common difference of the jacks; and $11\frac{1}{4}$ times 18 inches equals 16 feet $10\frac{1}{2}$ inches, the length for the corresponding hip or valley.

The cuts on the square are as follows: 12 and 6, seat and plumb cut of the common and jack rafters; 17 and 6, seat and plumb cut of the hip or valley; 12 on the tongue and $13\frac{5}{8}$ on the blade will give the side cut of the jack; they also give the face cut across roof boards to fit in the valley or over the hip, the blade giving

the cut in the former and the tongue in the
latter. The backing of the hip may also be
found by taking 18 on the tongue and 8 on the
blade, and the tongue will give the required
angle.

For an 8-inch rise, the lines from 12 and 17
(Fig. 54) would run to 8 on the blade, and their

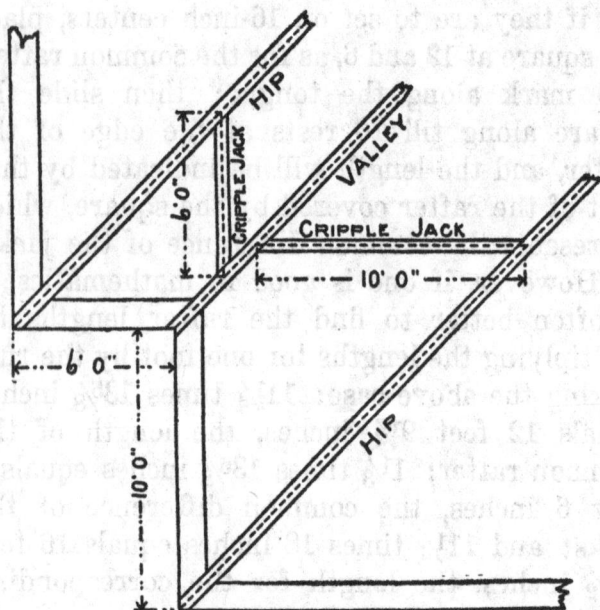

Fig. 56. Cripple Jack Rafters in Place.

lengths would consequently be changed; but the
formula remains the same.

To Find the Length of Cripple Jacks. The
length and cuts for a cripple jack can be found
just the same as for a jack resting against a
hip. The cuts of the cripple are the same at

both ends, and are identical with that for the upper end of a jack resting against a hip. Where the roof is all of the same pitch, the runs of the hips and valley will rest parallel with each other, as will be seen in Fig. 56. Now, here is a point that a great many do not grasp— namely, that **the run of the cripple jack is the same as the length of the plates that form the angle.** Thus, in the illustration, the length of the plate on one side is 6 feet, and on the other it is 10 feet, which represent the respective runs of the jacks in question. However, it should be remembered that this measurement is from center to center of hip and valley; and it is therefore necessary to make a deduction in the run equal to the thickness of the hip or valley. Or the length of the cripple may be found for the full run; then measure square back from the plumb cut the full thickness of the hip, which will be at the proper point for the plumb cut.

Framing Plan for Hip-and-Valley Roof. Fig. 57 is the plan of a common hip-and-valley roof, detailed to show all the different rafters, with their lengths and cuts, that usually enter into its construction. We shall assume it to be 10 inches to the foot. The view taken is from a point directly above. Consequently there is nothing in the plan to show what the rise is. In other words, if there were no pitch given the roof at all, the plan would show just the same, and the side cuts for hips and jacks would all be at an angle of 45 degrees, and their lengths would be as per the scale of the plan. That is, the first jack being placed 2 feet from the corner, its length would also be 2 feet; and

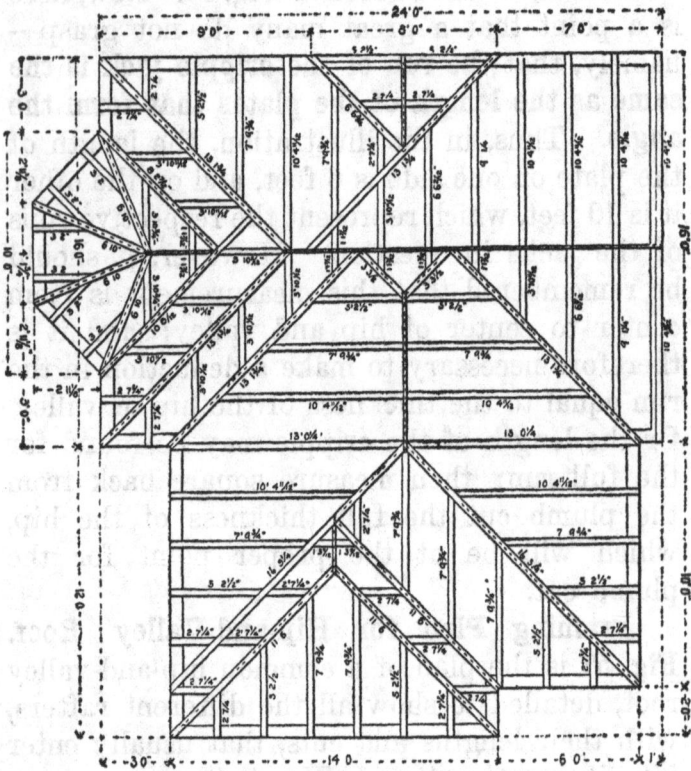

Fig. 57. Framing Plan for Hip-and-Valley Roof

these proportions taken on the square, as 12 and 12, will give what is generally called the **side cut**, but in reality should more properly be called the **top cut**. This, the reader will observe, is the regular miter, which is simple enough. Everybody understands so far; but when a pitch is given, this simple rule is usually forgotten.

In this example, the rafter, having a rise of 10 inches, has a gain of 7¼ inches in two feet; and this, added to its run, makes its length 2 feet 7¼ inches. Then the proportion of 2 feet and 2 feet 7¼ inches, taken on the square as 12 and 15$^{7}/_{12}$ inches, will give the cut. The side on which the larger number is taken, gives the cut. If the point of the jack is removed by cutting on a line parallel to the seat, it will be found that the angle of the cut is still 45 degrees, or just what the angle shows in the plan. The same rule applies for this as for the cut of the hip or valley. It also applies to the jack for an octagon, or any other corner.

In this example there is shown an octagon bay, and the side cut of the jack would be in the proportion of 1 foot and 3 feet 2 inches. The first, because that is the space that the foot of the jack is from the corner; and the latter represents the length of the jack. This jack, like all others, is simply part of a common rafter; reduced to a one-foot basis on the square, it is 5 and 15$^{7}/_{12}$ inches.

The lengths of the rafters are given from the edge of the plate to the center lines, as shown by the dotted line on the hips and valleys. Therefore, for the common rafters a reduction should be made for one-half the thickness of the ridge piece, by measuring square back that amount from the plumb cut.

It is not necessary to make any reduction for the jacks that rest on a plate, because the lengths given, if used for the long side, will make the jacks space all right, since the length is supposed to be taken along a line at the middle of the back. But a reduction equal to the diagonal of the thickness of the hip or valley should be made for the jacks that come in between a hip and valley. This also applies to the side cut of the hip where it rests against the ridge piece, by deducting half of the thickness of the diagonal of the piece. However, this is of small concern; and more than likely the variation, if not made, would go unnoticed.

In this example are shown some self-supporting hips and valleys, formed by letting one run by the other to a solid bearing. This is an important matter, which is too often overlooked; and consequently a sagged roof is the result.

There are other points about this plan that might be brought out. The figures to use on the square for a one-foot basis are as follows:

12 and 10—Seat and plumb cut of the common and jack rafters.

12 and 15 7/12—Side cut of the jack.

 5 and 15 7/12—Side cut of the jack.

13 and 10—Seat and plumb cut of octagon hip.

17 and 10—Seat and plumb cut of hip or valley.

17 and 19¾—Side cut of hip or valley.

The study of a rafter plan like this is valu-

able, for it contains practically all the elements
of any roof.

A Common Mistake. We wish to make a
small note right here which may set right some
of the younger members of the craft and a few
of the older heads who have never paid any attention to it. It is in the manner of adding the
projection for cornice. Fig. 58, at the left, shows
the wrong way, and makes the rafters too short,
causing the ridge joint to open as shown in the
middle part of the figure. The right way of getting this length is shown in the right-hand part
of the figure.

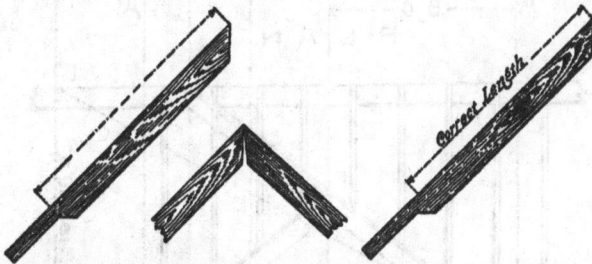

Fig. 58. Wrong and Right Methods of Figuring Rafter Lengths.

Framing for Roof Dormers. Fig. 59 represents the plan and the corresponding elevation
of the valleys in a roof dormer or gable. For example, 14 feet is taken for the run of the main
roof, and 8 feet for that of the gable. The roof
of the main part and that of the gable being of
the same pitch, it is evident that the ridge of the
latter will be below that of the former, as the rise
is proportional to the difference in the runs.

A-B represents the run of the long valley and **A' D** that of the short valley. Thus it will be seen that valleys framed in this way are self-

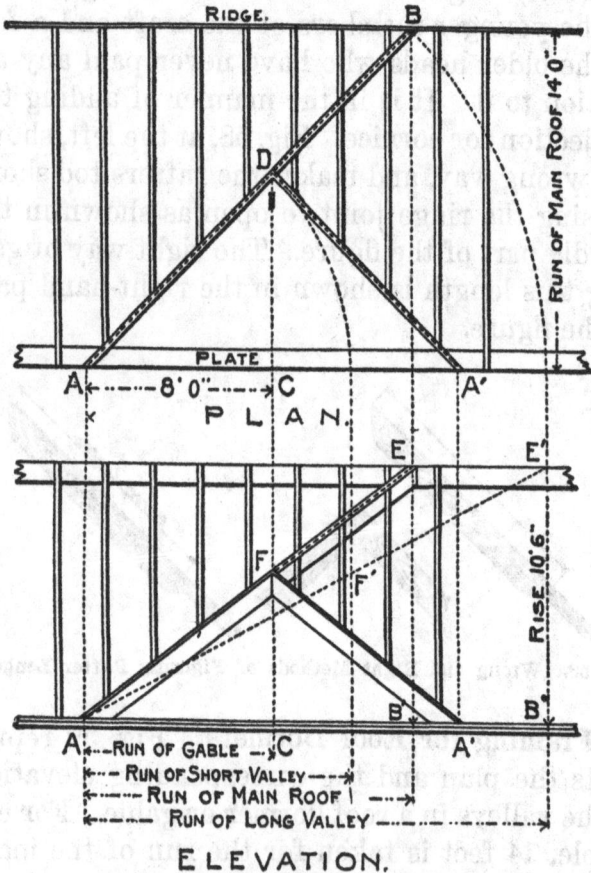

Fig. 59. Framing for Roof Dormers.

supporting. That part from **D** to **B** is what is generally termed **blind valley**, because it is concealed in the plane of the main roof. The meas-

urement should be taken along the center of the
back of the valley, as shown by the dotted lines;
and if backed—or, more properly speaking,
grooved, so that the roof boards will have a solid
bearing at all points—then the seat cut should
be made so as to bring the grooves in the plane
with that of the back of the common rafters.
This furnishes a problem in itself that is not so

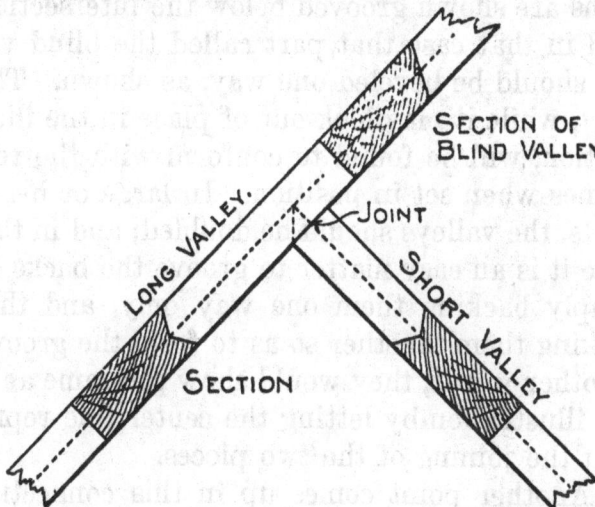

Fig. 60. Intersection of Valleys—Roof Dormers.

easily understood as may appear at first sight,
especially where there is a projection of the
rafter to form the cornice.

However, it is not usual to groove the valleys, as they are generally concealed from view
and otherwise not of enough advantage to warrant the extra work required. Where they are
not grooved, they should set proportionately

lower than the common rafter, so that the under
edge of the roof boards will intersect the center
of the back of the valley. Even then, that part
from **D** to **B** would have to be backed or beveled
on one side the same as for a hip, to bring the
center in plane with the common rafter.

Fig. 60 shows the plan of the valleys at the
intersection on a larger scale. In this the sections are shown grooved below the intersection;
and in that case that part called the blind valley should be beveled one way, as shown. This
part, while it may look out of place in the illustration, will be found to conform with the roof
planes when set in position. In large or heavy
roofs, the valleys should be doubled; and in that
case it is an easy matter to groove the backs by
simply backing them one way only, and then
spiking them together so as to form the groove.
In other words, they would show the same as in
the illustration by letting the center line represent the joining of the two pieces.

Another point comes up in this connection
that should not be overlooked before passing on,
and that is the joining of the short valley to the
long one. Simple as it is, builders sometimes
do not readily grasp that it is nothing more
than the plumb cut for the valley. It rests at
right angles from the long valley, and therefore
must rest square against it, just the same as if
against a level piece; and in this example, the
pitch being 3/8, 17 and 9 will give the cut.

Referring to the elevation part of Fig. 59,

the valleys are shown in position in the roof. They also show the same as the common rafters in their true position; but the valleys resting at an angle of 45 degrees from the common rafter, their lengths per scale are not easily arrived at without a few extra lines, which may be obtained as shown by the dotted lines from the plan to the elevation, as follows:

A-E represents the long valley in position from the point of sight, while **A-E′** shows its length. The same is true of the short valley. It is the same as **A-F** on the long valley. On a straight view, it represents the length of the common rafter for the gable, but its (the valley) length is found at **A-F′**.

Now we shall illustrate the above by simple lines on the steel square (see Fig. 61), using the same reference letters for the different parts, as shown in Fig. 59. The pitch being $3/8$, or 9-inch rise to the foot, we let 12 on the tongue of square No. 1 represent the starting point, and 9 on the blade the rise. The run of the main roof being 14 feet, measure back 14 inches along the line of the tongue and draw a line parallel to the blade to opposite 14 inches on that member, as at **B′ B**. The line from **A** to **B** will represent the run of the long valley. Now, by placing 17 on the tongue of square No. 2 at 12 on the square No. 1, and with the tongue along the line **A-B**, the heel will rest at 12 on square No. 1. Since the rise is 9 inches to the foot, a line from **A** passing at 9 on the square

No. 2 and intersecting the line **B-E′** (the rise
of the main roof) will represent the long valley;
and the line passing at 9 on square No. 1, intersecting the line **B′B** as at **E**, will represent
the common rafter for the main part.

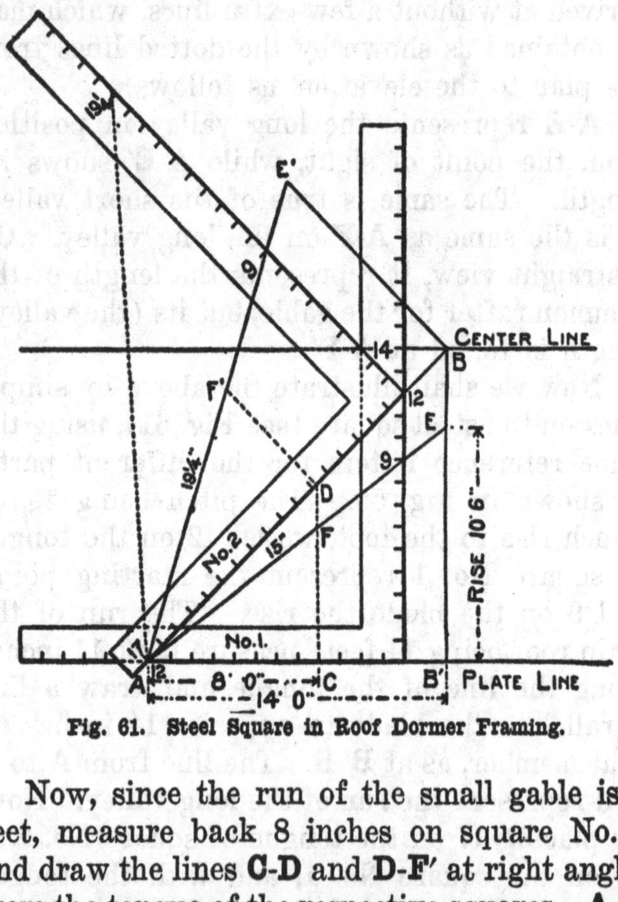

Fig. 61.　Steel Square in Roof Dormer Framing.

Now, since the run of the small gable is 8
feet, measure back 8 inches on square No. 1
and draw the lines **C-D** and **D-F′** at right angles
from the tongue of the respective squares. **A-F′**
will represent the short valley, and **A-F** the
corresponding common rafter to a scale of one

inch to the foot. The figures shown on the square intersected by the lines **A-E** and **A-E'** will give the seat and plumb cuts of the common and valley rafters respectively. The length of the diagonal lines on the squares are 19¼ and 15 inches, and these figures taken on the blade of the respective squares will give the side cuts for the valley and jack rafters.

In this illustration we have used two scales—the full scale on the steel square for a one-foot run, to obtain the cuts; and the $^1/_{12}$ scale, or one inch to the foot run, for the diagram of the roof, from which to obtain the length of the rafters. The fact that there are two scales employed may render the subject harder to grasp by some; but we trust that after a little study of this illustration will be clear.

The reader will observe that in all of our work we have adhered to 12 on the tongue as the starting point. We do this because it represents unity or the beginning, and therefore answers for any run or pitch given the roof. However, as a comparison, it might be well to illustrate this problem by the one-inch scale to the foot.

Bear in mind that while we illustrate these problems with two squares, only one is necessary, as the angles may be laid out with the different positions of the square and the required proportions taken on the same. As the run of the small gable is 8 feet, place the blade of square No. 2 at 8 on both the tongue

and blade (Fig. 62), with the heel opposite 14
of square No. 1 (because 14 represents the run
of the main roof). Now, since the rise is 9
inches to the foot, for 14 feet it would be 10
feet 6 inches. Then the line from 19¾ to 10½

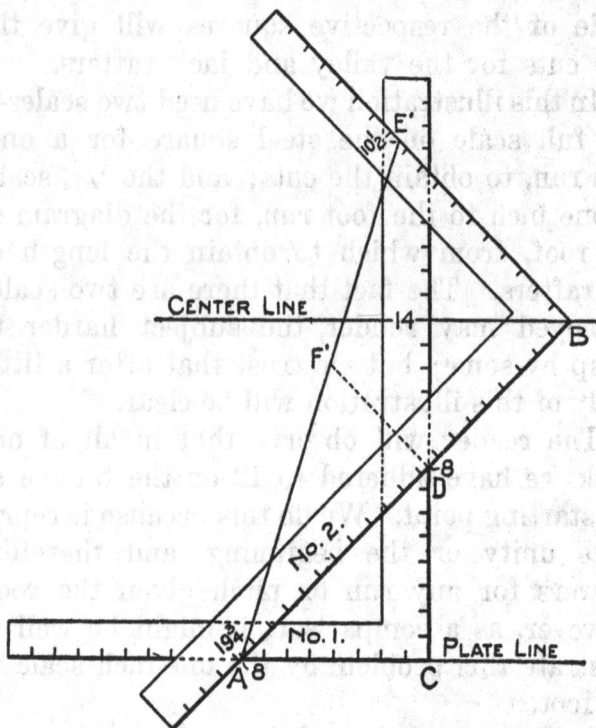

Fig. 62. Steel Square in Roof Dormer Framing.

will be the same length as **A E′** of like letters
in the previous illustrations. By drawing the
line **D-F′** at right angles to the blade, **A-F′** will
represent the length of the short valley.

As for the length of the common rafters, this is an easy matter to get by the scale method, by simply taking the run and rise of the roof on the tongue and blade and measuring diagonally across. However, while this does for working purposes for the more common run of work, it is not absolutely a correct method, because the least variation is magnified twelve-fold.

How to Roof a Circular Bay. The trouble that one carpenter had with such a job working "by rule o' thumb" is well illustrated in the following letter.

"I executed a piece of work last week which accidentally proved satisfactory to the contractor I am working for, but not to myself. Not knowing the proper way to go about it, I had a deuce of a time and worried myself half-sick.

"The job in question was framing a roof over a circular bay window, as they are termed here. The window had a radius of 15 feet, with a projection of 2 feet 6 inches and a rise at the center of 1 foot 6 inches. The roof was sheathed and covered with tin.

"The way I went about it at first was to strike my 15 feet radius, and then lay off the rise at center, allowing for projection. Next I struck the radius of my roof, which was 19 feet, something—do not call to mind just what it was. By doing so I expected to get the true rise and run of my rafters. The rafters were spaced on 32-inch centers—that is, I spiked my lookouts on every other joist, and set my rafters on top of them. I spaced off my drawing 32 inches, commencing at the center and working each way. Then took the run and rise of each rafter separately and cut them, but when I commenced to put them up something was wrong. The first, or center

rafter, was all right and the others were wrong. I felt
awfully cheap, but of course the work had to be done,
and not knowing any better way than by the old, old-
fashioned one, or the rule of 'thumb,' I finished my roof
that way. It looks all right, I suppose, to most people
outside of a good practical man, but I am not satisfied.
The chances are I will have more of the same work to do,
as they are getting to be quite the style.

"If it is not asking too much, I certainly would esteem
it a great favor if you would, or could, afford the time
to enlighten me as to the simplest method of executing
the above described work."

The above letter is printed in full, because
it is clear-cut and shows that the writer is after
information pure and simple. He admits his
error, and is anxious to avoid a like occurrence
in the future. His frankness in the matter
shows that he is on the right track to better
fit himself for his work and give value received
to his employer, with a good share of interest
thrown in.

We are all given to mistakes; and from one
another's experience we gain our knowledge
for paving the way for still others to follow.
To the experienced, such questions may seem
simple and a waste of words in the way of
explanation, but all such should remember that
they too were once groping in the dark for
information just as thousands of others are
to-day; and so it will ever be.

The great trouble experienced by the would-
be learner is that he does not stop to think—

that is, of the rules of application or relation
of one part to another.

Take, for instance, the above example. If
the bay window had been a full half-circle, this
man would without hesitation have framed his

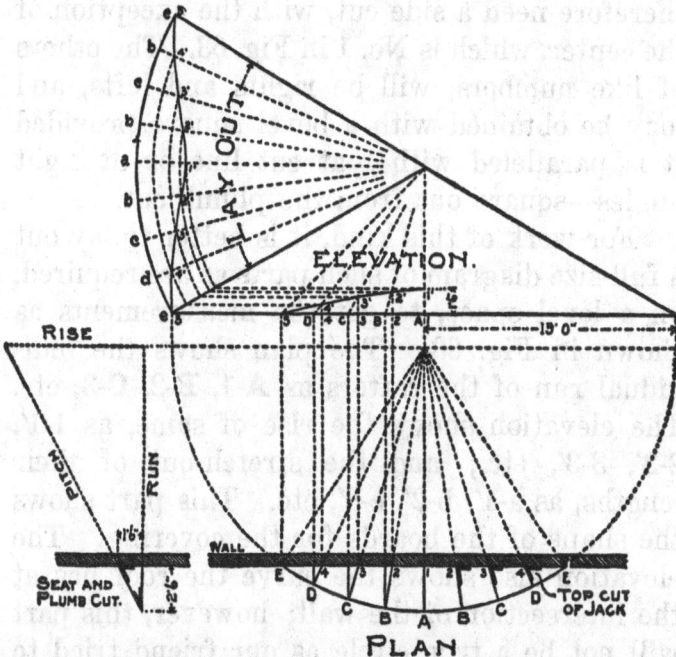

Fig. 63. How to Roof a Circular Bay.

rafters the same as for a circular roof—all of
same length and radiating from a common
center; but when only a fraction of the roof is
wanted, he forgets that the rafters must lie in
the same position as for the half-circle bay. In
one case, the wall line of the house, as it were,

cuts through the center of a circular conical roof; while in the other, it cuts off only the edge of the roof. Therefore, being a part of the same roof, the seat and plumb cuts must be the same in either case; but not being whole rafters, they must be considered as jacks, and therefore need a side cut, with the exception of the center, which is No. 1 in Fig. 63. The others of like numbers, will be rights and lefts, and may be obtained with a bevel square provided it is paralleled with seat cut line or at right angles—square out from the plumb cut.

For work of this kind, it is better to lay out a full-size diagram of such parts as are required, on a level space, to get the measurements as shown in Fig. 63. The plan shows the individual run of the rafters as **A**-1, **B**-2, **C**-3, etc. The elevation shows the rise of same, as 1-1′, 2-2′, 3-3′, etc.; and the stretch-out of their lengths, as a-1′, b-2′, c-3′, etc. This part shows the shape of the boards for the covering. The elevation also shows the curve the roof has at the intersection of the wall; however, this part will not be a true circle as our friend tried to have it. This should cause no worry because it will take care of itself, provided the rafters are cut to the right length and placed radiating to a common center. Further explanation, we trust, is not necessary.

To Develop Curved Rafters. A question that frequently comes up for solution, in getting out rafters that have a sweep or curve, is how to make the hips without scribing them from the common rafter, and make them so that all parts will line up properly. The method is shown in Fig. 64, which is self-explanatory. The curve for the common rafter can be anything desired, and should be laid off full size.

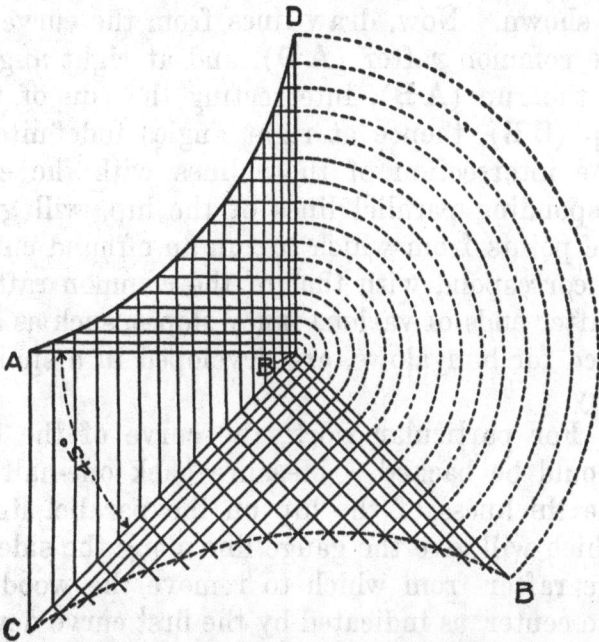

Fig. 64. Developing Curved Hip Rafter.

In the illustration, **A B** represents the run of the curve for the common rafter, and **C B** the same for the hip. **B D** represents the rise; and it necessarily follows that **A D** represents the curve given the common rafter. For a square-cornered building and where the pitch is the same on both sides, the run of the hip will necessarily rest at 45 degrees from that of the common rafter as shown.

Next, lay off any number of lines parallel to the run of the common rafter; also a like number with corresponding spacing parallel to the run of the hip, but to be of indefinite lengths, as shown. Now, draw lines from the curve of the common rafter (**AD**), and at right angles to the run (**AB**), intersecting the run of the hip (**CB**), thence at right angles indefinitely. The intersection of these lines with the corresponding parallel lines of the hip, will give the points from which to run an offhand curve to correspond with that of the common rafter. Rafter ends of various fancy slopes, such as are used for bungalows, are developed in a similar way.

For particular work, the curve of the hip should be backed. Measure back one-half of the thickness of the hip on the parallel lines, which will give the gauge line along the side of the rafter from which to remove the wood to the center as indicated by the first curve developed.

How to Use the Octagon Scale. On the tongue of almost all steel squares, there is a row of dots enclosed between two lines, and figured to a scale of tenths. This scale is called the **octagon scale,** and is designed for changing a square timber to an octagon, or for finding the width of the side of an octagon of a given

diameter. In Fig. 65 is shown a part of this
scale. But very few understand it, or are even
interested enough in it to look it up. It is
quite evident that the inventor did not understand the use of the plain steel square with its
standard scale of measurement, which is suffi-

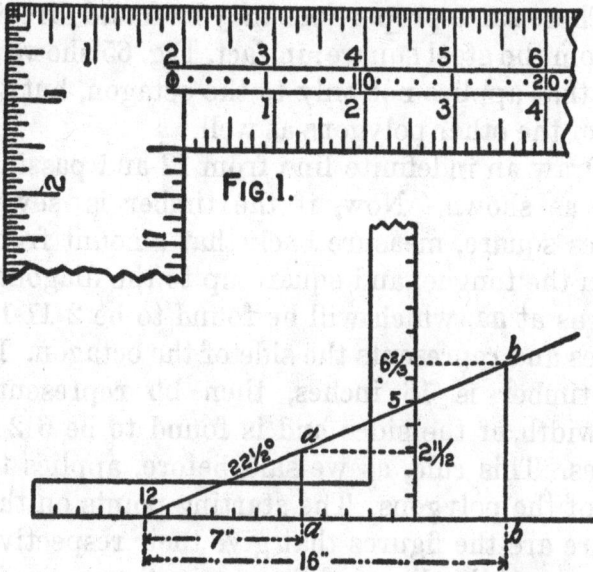

Fig. 65. Octagon Scale and Another Method.

cient for solving all problems of this kind, for
he wandered from the path of simplicity into a
byway to exemplify a single problem in the
polygons, then leaving the would-be learner
ignorant of any apparent reason why his scale
gives correct results. The solution is as follows:

Suppose it is desired to change a seven-inch

square stick to an octagon. Lay off a center line on all four faces of the timber; and from either side of this line set off a space equal to seven of the spaces shown on the steel square, which will be the point for the gauge line, from which to remove the wood at the corners to form the octagon.

These same proportions may be found directly from the steel square; in fact, Fig. 65 shows a rule that applies not only to the octagon, but to any of the other polygons as well.

Draw an indefinite line from 12 and passing at 5 as shown. Now, if the timber is seven inches square, measure back that amount from 12 on the tongue, and square up to the diagonal line, as at **aa**, which will be found to be 2 11-12 inches and represents the side of the octagon. If the timber is 16 inches, then **bb** represents the width of the sides and is found to be 6 2-3 inches. This rule, as we said before, applies to any of the polygons. The starting points on the square are the figures that give their respective miters, and the diagonal line across the square is governed accordingly.

How to Frame by Degrees With the Steel Square. The steel squares that are in general use do not contain a degree scale for framing purposes, though unquestionably it would be a good thing instead of some of the obsolete rules that now encumber its faces. However, in the absence of a protractor, the angle may be found as shown by the accompanying illustration, Fig. 66, as follows:

Lay off a line, as at **AB**, and apply the square as shown, and draw line **AC** from 12 on the tongue and passing at 12 on the blade. This line will be at an angle of 45 degrees to **AB**. Now, with the compass, strike an arc of any ra-

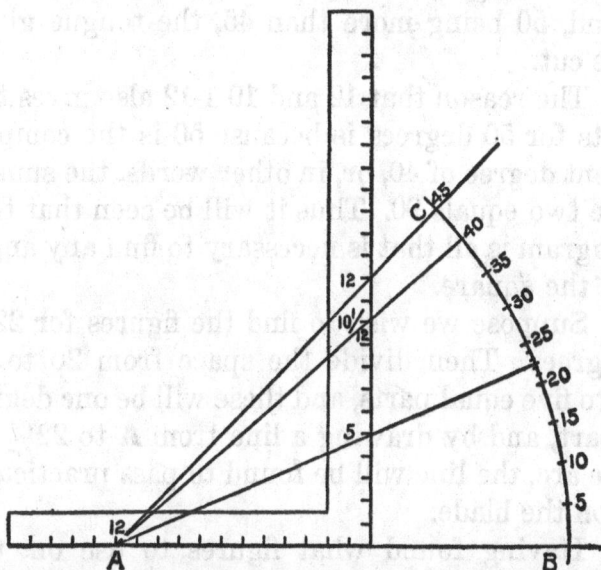

Fig. 66.   To Lay Out an Angle with the Steel Square.

dius—the larger the better, for the more accurate will be the final result. Divide this arc into nine equal parts, and these divisions will be five degrees apart, as shown by the figures opposite the divisions. Now, to find the figures on the blade for the degrees wanted, as 40, draw a line from **A** to 40 on the arc, and the line will pass at

10.06, or practically 10 1-12 inches on the blade, which will represent the figures to use on that member for either 40 or 50 degrees, the blade giving the plumb cut in the former and the tongue in the latter.

Vice versa for the seat cut. The changing point is at 45 degrees, and as 40 is less than 45, the blade gives the plumb cut. On the other hand, 50 being more than 45, the tongue gives the cut.

The reason that 12 and 10 1-12 also gives the cuts for 50 degrees is because 50 is the complement degree of 40, or, in other words, the sum of the two equals 90. Thus it will be seen that this diagram is all that is necessary to find any angle on the square.

Suppose we wish to find the figures for $22\frac{1}{2}$ degrees. Then divide the space from 20 to 25 into five equal parts, and these will be one degree apart, and by drawing a line from **A** to $22\frac{1}{2}$ on the arc, the line will be found to pass practically 5 on the blade.

Having found what figures to use on the blade for any desired degree, the procedure in roof framing is the same as in framing by the proportion of the span or per inch rise to the foot in run of the common rafter.

To Prevent a Ridge from Sagging. A method of construction such as is illustrated in Fig. 67 is at times desirable. It keeps the ridge from sagging, or plates from bulging out on the sides. The board, 1 by 8 inch, nailed on under

side of rafter, will prevent roof from sagging.
No collar beams are required. The board extends from end of plate at corner of building
diagonally to center of ridge.

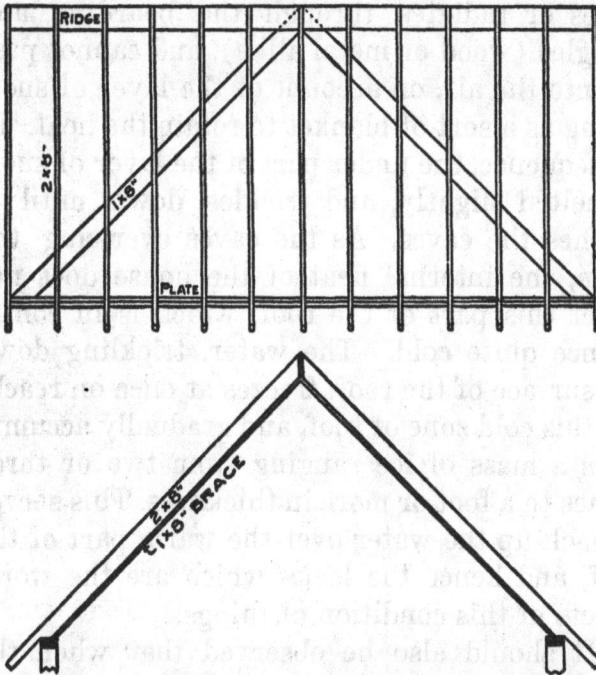

Fig. 67.   Good Brace for Ridge.

**Roof Construction to Prevent Formation of
Ice on Eaves.** This is a most important matter
in many parts of the country, especially in the
North.

Before suggesting a method of construction,
it may be well to examine into the causes of this

very prevalent trouble, which annoys many a householder winter after winter.

When a body of snow several inches thick lies on a roof, it forms a very effective non-conductor of heat. The warmth of the attic penetrates or radiates through the boarding and shingles (wood or metal alike), and cannot pass off into the air, on account of the layer of snow acting as a sort of blanket to retain the heat. In consequence, the under part of the layer of snow is melted slightly, and trickles down until it reaches the eaves. As the eaves overhang the walls, the internal heat of the house does not affect this part of the roof, which is in consequence quite cold. The water, trickling down the surface of the roof, freezes at once on reaching this cold zone of roof, and gradually accumulates a mass of ice ranging from two or three inches to a foot or more in thickness. This serves to back up the water over the warm part of the roof, and hence the leaks which are the worst effects of this condition of things.

It should also be observed that when the snow melts from the outside—that is, from the heat of the sun—no trouble occurs at the eaves, the resulting water running freely down off the roof. Seeing, then, that the cause of the trouble is the radiation, through the roof itself, of the internal heat of the house, the remedy evidently lies in preventing such radiation.

This may be done in two ways, either of which is fairly effective alone; but for first-class work and to insure the very best results it would be well to use both methods in combination.

The first method consists in thoroughly **deadening** the ceilings of all the upper rooms of the building—that is, to form a dead air space through which the heat of the rooms cannot escape. It cannot be too widely known by builders throughout the colder regions of this continent, that a lath and plaster ceiling allows a tremendous lot of heat to escape into the attic of a building; and, if for no other reason than the saving of fuel, this should be prevented. An effective method is to lay rough boarding on fillets near the upper edge of the ceiling joists, and to cover the same with some composition to render it air-proof. Fig. 68 shows the details of this method, which calls for slightly deeper joists to carry the extra weight of the deadening.

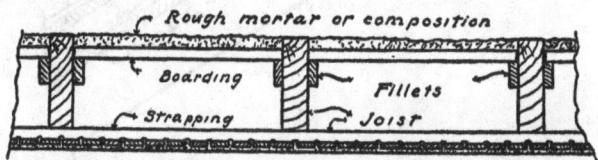

Fig. 68. Ceiling Construction to Prevent Heat Losses.

There are, of course, several **pugging** compositions sold for deadening purposes, but an entirely effective one is a mixture of coarse mill sawdust and plaster of Paris. The sawdust and plaster are mixed together and made into a mortar with water in the ordinary way, and a coating laid on the boarding, about one inch in depth.

Care must be taken that the spaces between the
ends of the joists over the wall-plates are properly filled, either by lath and plaster or by a
piece of board cut tightly in between and nailed
in position. This is important, for the effective-
ness of any dead air space as a non-conductor of
heat (or cold) depends upon its being absolutely
tight everywhere.

Many buildings, such as churches, halls, and
schools, have no attics, but are open to the roof

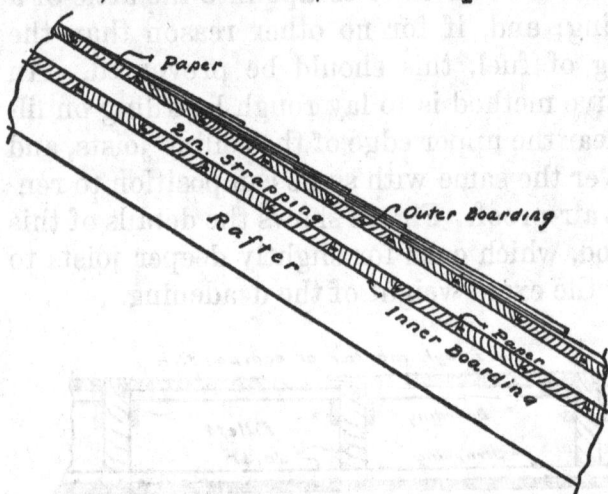

Fig. 69. Roof Construction to Prevent Heat Losses.

timbers, and obviously the foregoing would not
apply in such cases. For the prevention of heat
radiation in roofs of this class, a double roof is
the only effective remedy. A detail of this is
shown in Fig. 69, from which it will be seen that
the sheathing is laid on the rafters in the usual
way, and then covered with good building paper.

Two-inch strapping is then applied, and upon
this a second layer of sheathing is laid. This is
covered with paper and shingles in the regular
manner, thus forming a dead air space as desired. Such an arrangement will also prevent
the condensation of moisture upon the inside of
the roof, and the consequent annoyance from
water dropping, which is so often experienced
in churches and halls when well-warmed inside
during zero temperatures out of doors.

As remarked earlier in this discussion, to be
absolutely sure of preventing the formation of
ice near the eaves, it would be well to adopt both
the deadening of the ceilings and the doubling
of the roof. It will be found, however, that the
thorough deadening of the ceilings will generally
be sufficient; and very few architects specify
both methods, except in the most expensive
structures.

Stair=Building Simplified

Rules for Proportioning. Until about the time of Queen Elizabeth, the staircase, now so important a feature in all houses, was of small note. Previously, stairs were built in every case on a circular plan, revolving around a central axis or newel. These were known as **turret** or **corkscrew stairs.** Stairs with wide, straight flights were introduced during the sixteenth and seventeenth centuries, and were made the leading feature in the mansions of the Elizabethan style. They were very massive in design, with heavy oak balusters and enormous carved newels with ornamental panels. Many staircases of similar design but of lighter construction now exist in England, many of the more modern ones having cast-iron railings. In most houses built to-day the stairway is a most important feature, nevertheless its construction is frequently, and in fact usually, left with little thought as to design or convenience.

In planning stairways, care should be taken to have sufficient room so that the height of the riser is not too great. The distance from floor to floor in inches should be divided into a certain number of vertical distances, each of which is termed the **rise,** and which is usually from 7 to 8 inches. A rule frequently applied in proportioning the rise to the width of the **tread** is, that the rise in inches multiplied by the run or tread in inches shall be about 70 or 75. According to this, a 7-inch riser will call for a 10½-inch tread. The workman will readily see from this rule that the greater the rise the less the run, and the less the rise the greater the run, the proportions varying to suit different conditions.

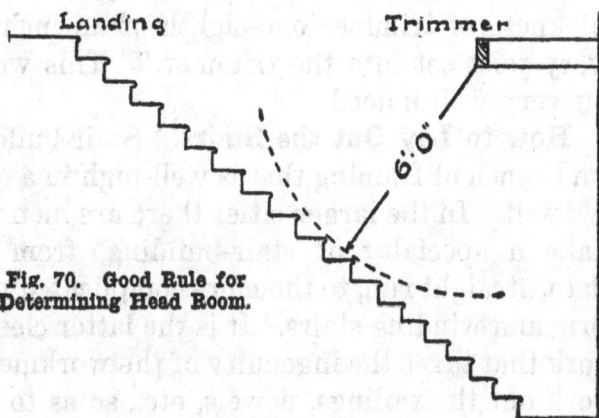

Fig. 70. Good Rule for Determining Head Room.

Another rule is based on the fact that an easy pace on the level is 23 inches. In going upstairs, however, a person passes upward as well as forward at the same time, and the rule is that "twice the rise, plus the tread, should equal 23." This gives very good results in practice.

Head Room. A good rule for getting proper head room in a flight of stairs is to count **down** 13 or 14 steps from the top, and plumb up for the face of the trimmer. This gives good clearance for a tall man.

Nicholson's famous old Scotch work on building construction (1805) gives another good rule, which is illustrated in Fig. 70. With the bottom front edge of the trimmer as center, and a radius of 6 feet, describe an arc of a circle, from which the nosing line must be kept clear to give good head room under the trimmer.

Nicholson's rule for finding the thickness of trimmers is also useful. It was "to add to the thickness of trimmer one-eighth of an inch for every joist set into the trimmer." This works out very well indeed.

How to Lay Out the Stairs. Stair-building is a branch of framing that is well-nigh in a class by itself. In the larger cities there are men who make a specialty of stair-building, from the plain, straight run, to the more complicated platform and winding stairs. It is the latter class of work that taxes the ingenuity of the workmen to work out the railings, newels, etc., so as to rest in the proper planes with the pitch given the stairs. To do this successfully, requires one well up in detail drafting and having a practical knowledge of geometry to lay out the turns and easements on the rough timbers, so as to be able to work them out in the finished product. However, the old-fashioned winding stair, with its ever winding rail, is no longer considered, from an architectural standpoint, to add to the beauty of the interior of the house, to say nothing of the added expense that must necessarily follow in the construction of circular stair work. Of

course, there are times where winding stairs
conform to certain space better than stairs with
square turns; but even then, in most cases, it
would be better to allow room for a good easy
stair with square turns, and make the rooms and
hallways conform to it.

So the question is how to lay out the necessary room for a comfortable and convenient
stair.

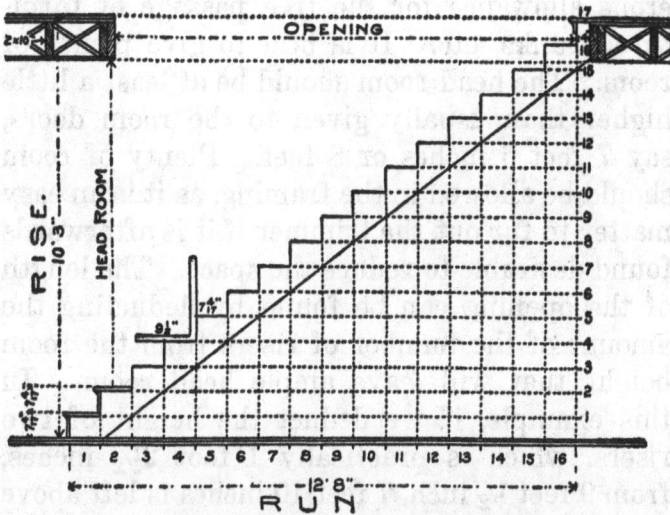

Fig. 71. Layout of Simple Stairs.

Lay out the stairs first, allowing ample
space; and plan the rooms accordingly. Fig. 71
shows a straight run of stairs which may be
taken as typical; it is the simplest of any to
build. From the starting point to the level of
the floor above, is 10 feet 3 inches, or 123 inches;

that there are to be risers. The pole can also be used to advantage where the total run is desired to come inside a given space, as it obviates the necessity of a mathematical problem that usually runs into fractions.

For general use, however, in calculating the

Number of Risers	RISE OR RUN.																
1	6½	6¾	7	7⅛	7¼	7⅜	7½	7⅝	7¾	7⅞	8	8¼	8½	9	9½	10	10½	11
2	11	11½	12	12¼	1'2½	12¾	1'3	1'3¼	1 3½	1 3¾	1'4	1'4½	1'5	1'6	17	1'8	1'9	1'10
3	1'7½	18¼	1 9	19⅜	1 9¾	1 10⅛	1 10½	1 10⅞	1 11¼	1'11⅝	2'0	2'0¾	2 1½	2 3	2'4½	2 6	2'7½	2'9
4	2'2	2'3	2'4	2'4½	2 5	2 5½	2 6	2 6½	2 7	2 7½	2'8	2 9	2 10	3 0	3 2	3 4	3 6	3 8
5	2 8½	2 9¾	2 11	2 11⅝	3 0¼	3'0⅞	3'1½	3 2⅛	3 2¾	3'3⅜	3 4	3'5¼	3'6½	3'9	3 11½	4 2	4'4½	4'7
6	3 3	3'4½	3'6	3'6¾	3 7½	3 8¼	3 9	3 9¾	3'10½	3 11¼	4'0	4'1½	4 3	4'6	4 9	5'0	5'3	5'6
7	3 9½	3'11¼	4'1	4 1⅞	4 2¾	4'3⅝	4 4½	4'5⅜	4'6¼	4 7⅛	4'8	4'9¾	4 11½	5'3	5 6½	5'10	6' 1½	6'5
8	4 4	4'6	4'8	4 9	4 10	4 11	5 0	5'1	5'2	5'3	5'4	5 6	5 8	6'0	6'4	6'8	7'0	7'4
9	4 10½	5'0¾	5'3	5 4⅛	5 5¼	5 6⅜	5 7½	5'8⅝	5 9¾	5'10⅞	6'0	6 2¼	6'4½	6'9	7 1½	7'6	7'10½	8'3
10	5 5	5 7½	5'10	5 11¼	6 0½	6 1¾	6 3	6'4¼	6 5½	6'6¾	6'8	6 10½	7'1	7'6	7 11	8'4	8'9	9'2
11	5 11½	6'2¼	6'5	6'6⅝	6 7¾	6 9⅛	6 10½	6'11⅞	7'1¼	7'2⅝	7'4	7 6¾	7'9½	8'3	8 8½	9'2	9'7½	10'1
12	6 6	6'9	7'0	7'1½	7'3	7 4½	7'6	7'7½	7 9	7'10½	8'0	8'3	8'6	9'0	9'6	10'0	10'6	11'0
13	7'0½	7'3¾	7 7	7'8⅝	7'10¼	7'11⅞	8 1½	8 3⅛	8 4¾	8'6⅜	8'8	8 11¼	9 2½	9'9	10 3½	10'10	11'4½	11'1
14	7 7	7'10½	8 2	8 3¾	8 5½	8'7¼	8 9	8 10¾	9'0½	9'2¼	9'4	9 7½	9'11	10'6	11'1	11'8	12'3	12'10
15	8 1½	8 5¼	8'9	8 10⅞	9 0¾	9 2⅝	9'4½	9 6⅜	9'8¼	9'10⅛	10'0	10 3¾	10 7½	11'3	11 10½	12'6	13'1½	13'9
16	8'8	9'0	9'4	9'6	9'8	9 10	10'0	10 2	10 4	10'6	10'8	11'0	11'4	12'0	12'8	13'4	14'0	14'8
17	9'2½	9'6¾	9'11	10 1⅛	10'3¼	10 5⅜	10 7½	10 9⅝	10 11¾	11'1⅞	11 4	11'8¼	12'0½	12'9	13'5½	14'2	14 10½	15 7
18	9'9	10'1½	10'6	10'8¼	10 10½	11'0¾	11 3	11'5¼	11'7½	11'9¾	12'0	12'4½	12 9	13'6	14'3	15'0	15'9	16'6
19	10'3½	10 8¼	11'1	11'3⅜	11 5¾	11'8⅛	11 10½	12 0⅞	12 3¼	12 5⅝	12 8	13 0¾	13'5½	14'3	15'0½	15'10	16'7½	17'5
20	10'10	11'3	11'8	11'10½	12'1	12 3½	12 6	12 8½	12'11	13'1½	13'4	13 9	14'2	15'0	15'10	16'8	17'6	18 4
21	11'4½	11 9¾	12'3	12 5⅝	12'8¼	12'10⅞	13 1½	13'4⅛	13 6¾	13 9⅜	14 0	14 5¼	14'10½	15 9	16'7½	17'6	18'4½	19'3
22	11 11	12 4½	12 10	13 0¾	13'3½	13'6¼	13 9	13 11¾	14 2½	14'5¼	14'8	15 1½	15'7	16'6	17'5	18'4	19'3	20'2
23	12'5½	12 11¼	13'5	13'7⅞	13 10¼	14'1⅝	14'4½	14'7⅜	14'10¼	15'1⅛	15 4	15'9¾	16'3½	17'3	18'2½	19'2	20'1½	21'1
24	13'0	13'6	14 0	14'3	14 6	14'9	15 0	15 3	15 6	15'9	16'0	16'6	17'0	18'0	19'0	20'0	21'0	22'0
25	13 6½	14 0¾	14 7	14'10⅛	15'1¼	15 4⅜	15 7½	15 10⅝	16 1¾	16 4⅞	16'8	17'2¼	17'8½	18 9	19'9½	20'10	21'10½	22'11
26	14'1	14 7½	15'2	15 5¼	15 8½	15 11¾	16 3	16'6¼	16 9½	17 0¾	17 4	17'10½	18 5	19 6	20 7	21'8	22'9	23 10
27	14 7½	15'2¼	15 9	16 0⅜	16 3¾	16 7⅛	16 10½	17 1⅞	17'3¼	17 8⅝	18 0	18 6¾	19'1½	20'3	21 4½	22 6	23 7½	24'9

Fig. 72. Table Giving Number of Treads or Risers of Any Width for Any Size Space.

layout for stairs, the accompanying table, Fig. 72, will be found very handy and useful as a time-saver. The first row of figures running down the left-hand side, represents the number of risers, while the first row running across the top represents either the rise or the width of the tread. Those in the following lines repreis, it is not confined to a certain space like the rise from floor to floor. Therefore, a few inches in the run of a straight flight of stairs does not usually make any difference, thus leaving it to the builder to select at once the width of tread desired. When this cannot be done, then the allotted space must be arrived at in the same manner as that given in the above for the risers. But after all, it should be remembered that while the measurements can be accurately found by the aid of this table, its greatest utility is as a quick reckoner, in laying out the space and proper openings for the finished stair work. In that case it is not necessary to calculate down to the minuteness required in the building of the stairs.

Types of Stair Construction.

Taking up the various details of construction, the **housed string stair,** one of the simplest and at the same time an important type, presents itself. This class of stairs may be divided into two kinds—first, where the stair is between walls (that is, where both strings are fastened to and supported by the walls); and second, where only one of the strings is fastened to the wall, and the other (the face or outside string) is free.

The first is the cheaper, and is used very much in small cottages, and also as a rear stair in the better grade of houses. Of course, very often both of these stairs are framed without the housed string. The treads are carried on a rough string, and the finished string is fastened to the treads and risers by nailing through it into the treads and risers; but this is very poor construction.

By the term **housed string** is meant a string notched out to receive the ends of the treads

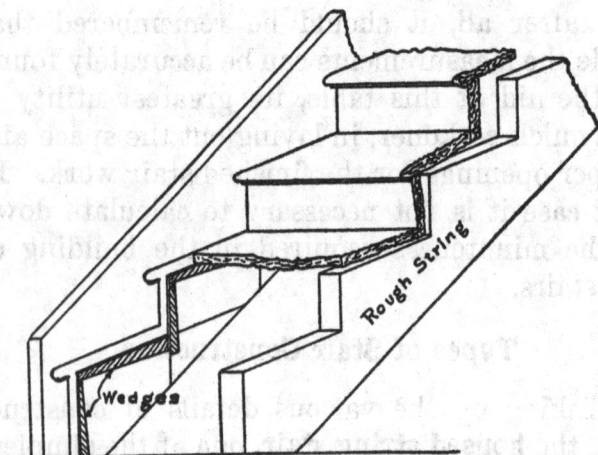

Fig. 73. Housed-String Stair Construction.

and risers. An examination of Fig. 73 will show clearly what is meant.

In the stair between two walls, rough strings are unnecessary, unless the stair is over 2 feet 6 inches wide, when a rough string must be provided under the middle of the stair. The finished strings are fastened to the walls, and are more rigid than if a rough string were the means of support.

Laying Out Stair Strings. After determining the tread and riser lengths, proceed to lay out the string. A little device very helpful in laying out a string is a **gauge-board**, as shown in Fig. 74, upon which has been cut the proper length of tread and riser to the pitch of the stair. In notching out the treads and risers, the notches should be cut large enough to receive a small wedge below the tread and back of the riser. These are used to make a tight

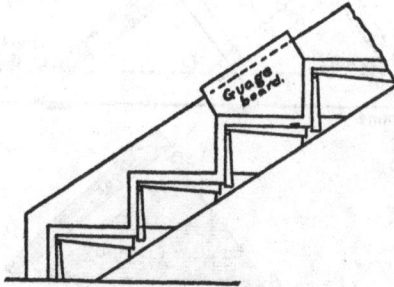

Fig. 74. Use of Gauge-Board.

fit in front, where the treads and risers come against the edge of the notches. When the stairs are put together, the wedges are covered with glue before being driven into place.

Another method of laying out housing for stair strings is as follows:

Joint top edge of string-board straight; draw a gauge-line down the required distance for the center of nosings. Then, having found the rise and run of riser and step, set dividers from rise on blade of square, and step off on gauge-

line the required number of treads. Next take
a center bit the size of nosings, and start holes
at these points. Next mark a **pitch-board,** as
shown in the lower part of Fig. 75, at **D,** and a
wedge-shaped stick **F,** the thickness of tread,
plus the shape of wedge to be used in gluing
up the stairs. Now place pitch-board on string,
as shown at **D** in the upper part of Fig. 75,

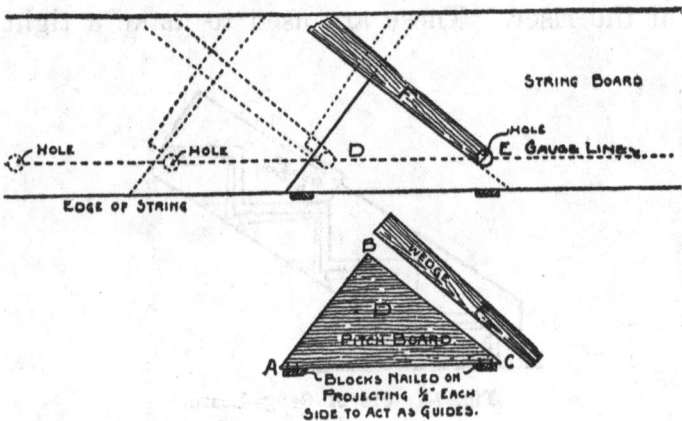

Fig. 75. Method of Laying Out Stair Strings.

and slide it along edge of string-board until edge
touches hole **E.** Mark tread and riser; and,
before moving pitch-board, place the wedge
stick as shown at **F** and at dotted lines, and
mark outside. Proceed to next tread in same
manner, and so on. The projection of nosing
may be regulated by making pitch-board longer
or shorter on line **B-C;** but do not change the
pitch.

Closed-String Stairs. A **closed-string** or **box** stair—that is, one between two walls—is built up against one of the walls before the second wall is built, which, when the stair is in place, is set up against it, and the string nearest to this wall is then fastened to the studding. If the stair is put in and lathing done afterwards, pieces of inch stuff will have to be cut between the studding along the string to receive the ends of the lath. A better way is to lath wall No. 1 before building the stair; then put in the stair; and when placing the studding of wall No. 2, leave the thickness of a lath between the studs and the near string. Then lath this wall, shoving the lath through behind the string and nailing them below and above the stair. When this is done, the string may be fastened to the studding by nailing through the string beneath the treads and risers.

Care must be taken, however, that the string lies well against the studding—even if it is necessary to put in thin blocking strips at each nailing place. Otherwise, there is danger of breaking the glue joint, where the treads and risers fit into the string.

There are several methods of **joining the risers and treads.** In Fig. 76 are shown the various methods of doing this. Some are considered better than others, but that is much a matter of opinion.

At **A** is shown how the risers come down upon the tread, and the tongue on the front side

of the riser fits into a groove in the tread. If
the fit is not good, there will be a crack in the
front of the riser, which will become more
apparent as the stair becomes older.

At **G** is another method. This is the common way when there is a rough string under
the middle of the stair. At **f** is a similar
arrangement except that the riser goes down
behind the tread.

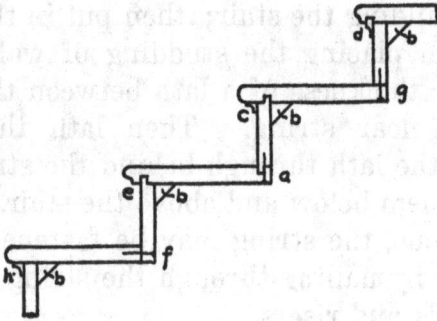

Fig. 76. Methods of Joining Risers and Treads.

All of these joints should be nailed or fastened with wood screws.

At **e** and **d**, Fig. 76, are shown two ways of
joining the tread and riser at the nosing. They
are practically the same, the one at **d** having
the moulding set into the groove; while at **e**
a tongue is cut on the front of the riser to fit
into the groove in the tread. In **d** and **e** there
is the danger of having the nosing break off,
because the tread, unless made of a rather thick
piece of wood, may crack over the groove.

In **c** and **h** are two more satisfactory methods. At **c** a tongue is cut on the back edge of the riser, and the groove in the tread is as a result farther back from the front of the tread than in **e** and **d**. A small moulding is put under the nosing, and there is little danger of the nosing breaking off. In **g** no groove is cut in the tread, so that the full strength of the tread is preserved. In this case, however, it is almost necessary to put a triangular strip, **b**, as shown, so as to fasten the tread and riser together. This strip may be put on with good results in all cases, as it will stiffen up the work considerably.

Open-String Stairs. Now taking up the stair which has one side open—that is, the string farthest from the wall, which is the face string— Fig. 77 shows a plan and side elevation of such a string. The treads and risers are housed the same as in the other stair, but the string must be made somewhat differently to fit the conditions.

W.S. shows the wall string; **R.S.** the rough string placed there to give the structure strength; and **O.S.** the outer or cut string. At **a,a,** the ends of the risers are shown; and it will be noticed that they are mitered against the vertical or riser line of the string, thus preventing the end wood of the riser from being seen. The other end of the riser is in the housing in the wall string. The outer end of the tread is also mitered at the nosing, and a piece of stuff made or worked like the nosing is mitered against, or returned at the end of the tread.

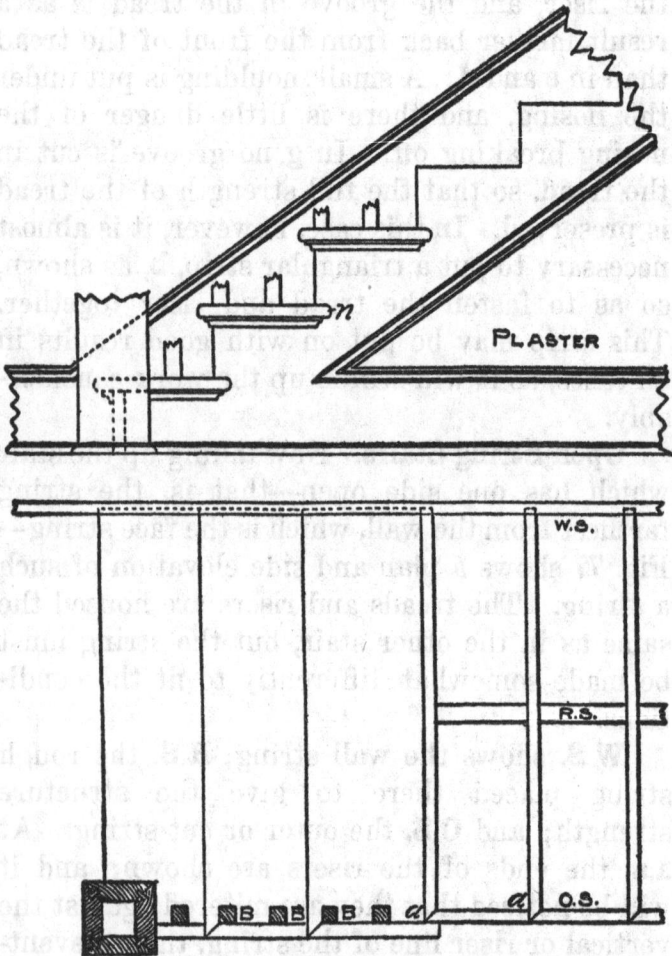

Fig. 77. Plan and Elevation of Open-String Stair.

The end of this returned piece is again returned on itself back to the string, as shown in the upper portion of the cut, at **n**. The moulding, which is a five-eighths cove in this case, is also returned round the string and into itself.

The mortises shown at the black points, **B,B,B**, etc., are for the balusters. It is always the proper thing to saw the ends of the tread ready for the balusters before they are attached to the string; then, when the time arrives to put up the rail, the back end of the mortise may be cut out, when the tread will be ready to receive the baluster. The mortise is dovetailed;

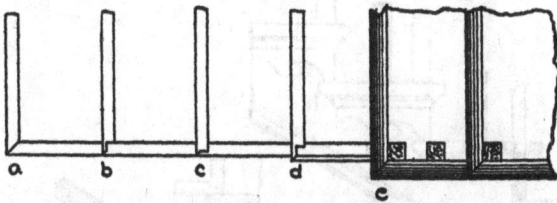

Fig. 78. Miter Joints for Open Stair Work.

and, of course, the tenon in the baluster must be made to suit. The tread is finished on the bench; and the return nosing is fitted to it, and tacked on so that it may be taken off to insert the balusters when the rail is being put in position.

In an open stair, and especially one in which the treads project over the face of the string, it is desirable to have the work rather well finished in order to present an attractive appearance, one that will harmonize with its surroundings. In the modern dwellings of to-day, the front hall or the stair hall is made larger than is necessary to accommodate merely the stair. The reception hall and stair hall are combined, and appropriately so; but it is necessary then to finish the room more elaborately than if it were used as a stair hall only. One thing to be provided for is that no end of any piece of wood shall show. In order to accomplish this in the riser, the rise in the string is mitered, and the end of the riser is cut on the same miter.

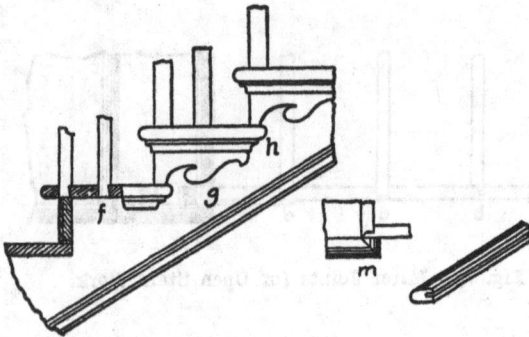

Fig. 79. Construction and Finish of Open Stairs.

In Fig. 78, the different ways of mitering are shown. At **a** is a miter of forty-five degrees cut on both the string and the riser. This is the simplest method and the one more often used because of the saving in time. At **b** the riser has a shoulder to fit against the string, and only the outside is mitered. This makes a more rigid joint. At **c** the miter is cut at the front as at **b**, and the string is cut out to receive the remainder of the riser. Here the riser gets a stronger bearing upon the string; while in **b**, only the front of the riser gets a bearing.

When it is desired to make the face of the string more ornamental, a thin bracket is placed against the string, as shown in Fig. 79, at **g**. When this is done, the riser must be longer than the thickness of the bracket. This is necessary because the bracket is mitered to the riser. The cove under the nosing is placed upon the bracket, just as it is returned upon the face of the string in the case where the bracket is not used. The lower front part of the bracket rests upon the returned nosing of the tread.

In the best grade of work the brackets are glued upon the string, but ordinarily they are nailed on with brads, which are then set and the holes filled with putty.

The return nosing is mitered at the front of the tread to fit the nosing over the riser. At the back of the tread, a return is cut as seen in Fig. 79, at **m**, which is a plan of **h**.

After the balusters are in place, the return nosing is nailed to its proper place, and the nail-holes filled with putty; or a groove may be cut into it, as shown at the right in Fig. 79. On the end of the tread, as seen in Fig. 80, at **a**, a similar groove is cut and a thin piece of wood or tongue glued into the groove in the end of the tread. This tongue should properly have the grain of the wood run in the same direction as the grain in the tread. The return nosing is then fastened into place by gluing the tongue and the groove, and driving the nosing to a tight fit.

Another way to fasten the nosing is to cut notches on the under side of the tread, and put wood screws through into the return nosing.

A glance at these figures will show how the balusters are dovetailed into the tread.   The

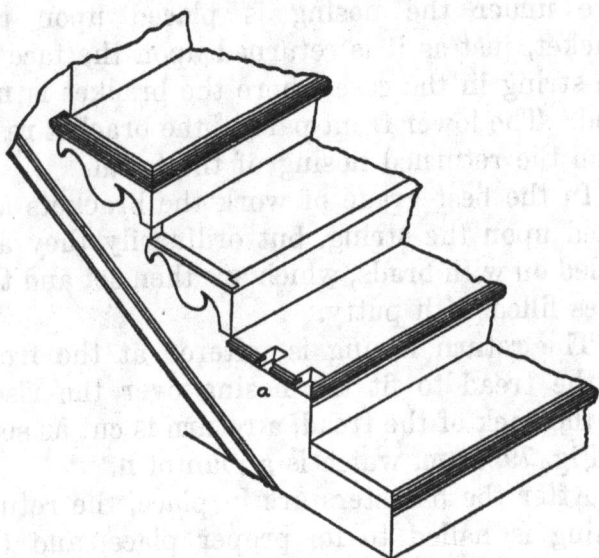

Fig. 80.   Construction and Finish of Open Stairs.

outside of the baluster should be flush with the face of the string; and where a bracket is used, this must be considered the face of the string.

Various Stair Arrangements.   There are many ways of building stairs; what may be suitable in one place, may not be in another.   Whatever style is wanted, however, should be care-

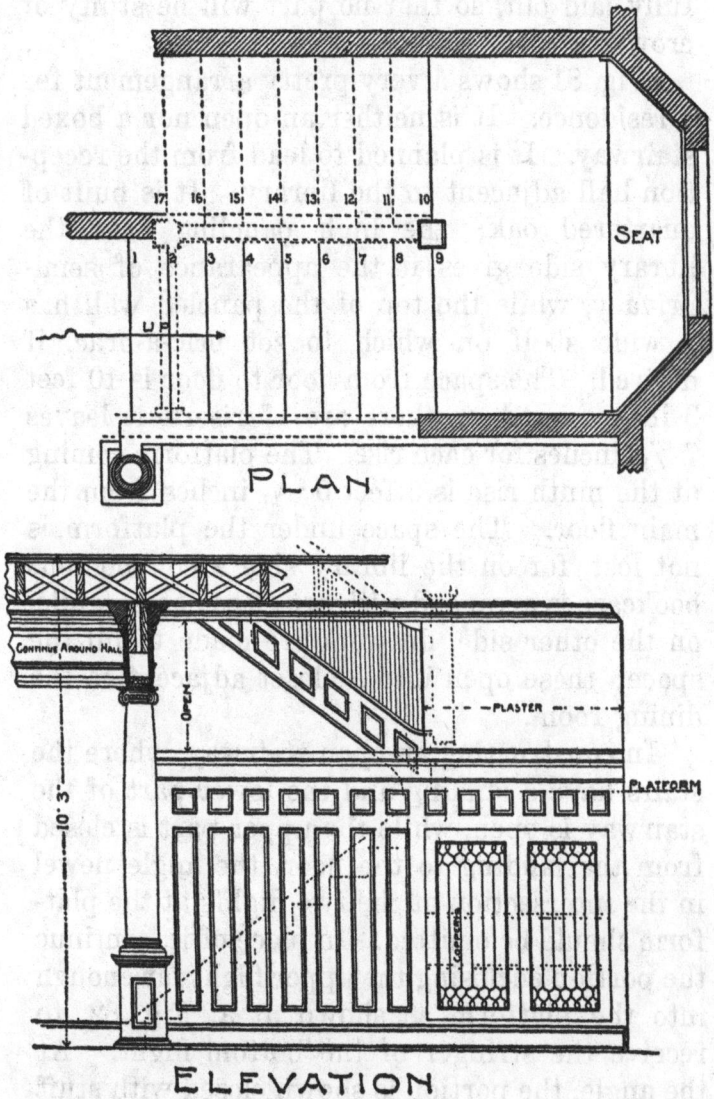

Fig. 81. Desirable Stair Arrangement for Residences.

fully laid out, so that no part will be stuffy or crowded.

Fig. 81 shows a very pretty arrangement for a residence. It is neither an open nor a boxed stairway. It is planned to lead from the reception hall adjacent to the library. It is built of quartered oak; the high paneling next the library side gives it the appearance of semi-privacy, while the top of the paneled wall has a wide shelf on which to set bric-a-brac, if desired. The space from floor to floor is 10 feet 3 inches; and, as there are 17 risers, it leaves $7\,^4/_{17}$ inches for each rise. The platform coming at the ninth rise is 5 feet $5\,^2/_{17}$ inches from the main floor. The space under the platform is not lost, for on the library side a commodious bookcase is arranged with art glass doors; while, on the other side, drawers are made to fill the space; these open into a closet adjacent to the dining room.

In constructing an open stairway, where the stairs have a landing and the lower part of the stairway is open, while the upper part is closed from the landing to the floor, the angle newel in the intersection of the two flights at the platform should be omitted. In place of it, continue the portion enclosing the upper flight far enough into the platform as shown at **a**, Fig. 82, to receive the stringer of the bottom flight. At the angle, the portion is shown closed with stuff equal in thickness to the thickness of the stringers; and the stringer of the closed upper

flight, which is on the inside of the portion, will
butt against the casing. The rail of the bottom
flight will be fastened to the casing above the
bottom stringer; while the rail for the upper
flight will have to be fastened on brackets to

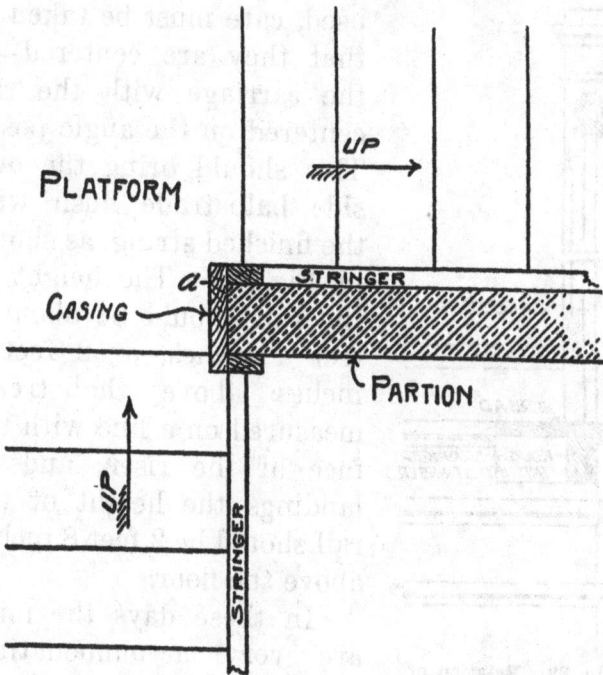

Fig. 82. Arrangement for Platform Stairs.

the side of the portion, and is known as a
wall rail.

Winders should never be used if it is possible
to avoid them, their great objection being the
narrowness of the tread along the line of travel,
which is a line generally taken about 14 inches

from the rail. Where they are absolutely necessary, it is better that the rail be made continuous with the ends of the treads in the form of a cylinder, and that the risers do not radiate from a common center.

Where angle posts are used, care must be taken so that they are centered on the carriage with the rail centered on the angle posts. This should bring the outside balustrade flush with the finished string, as shown in Fig. 83. The height of the rail should be about 2 feet 4 inches, or 2 feet 6 inches above the tread, measured on a line with the face of the riser; and on landings the height of the rail should be 2 feet 8 inches above the floor.

In these days the mills are very accommodating, and will, as a rule, work out any stair problems that the carpenter may have.

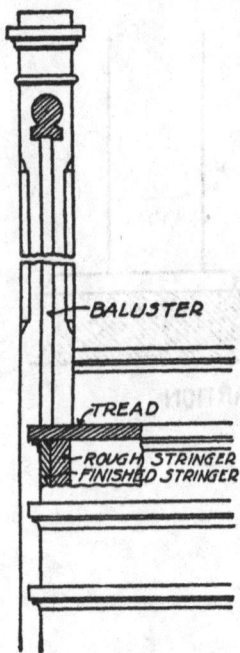

Fig. 83. Relation of Post, Rail, and String.

The principal thing for the carpenter to consider in planning a stair is to see that he has sufficient room to get the required number of steps and ample head room. If he is sure of these, the mill will usually do the rest of the designing for him.

In constructing the rough framework for the stairs, care must be taken to have the stringers sufficiently strong to carry the weight. They

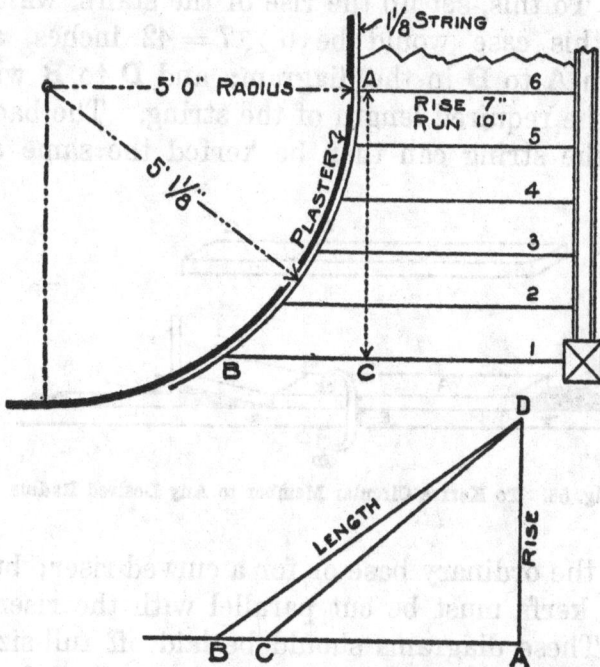

Fig. 84. Stair against a Circular Wall.

should never be less than 6 inches in the narrowest part. As a rule, it is best to put all of the rough framework for the stairs in place before the lathing and plastering is done.

Stair to Fit Circular Wall. It is sometimes

a problem to lay out the string to fit around
a circular wall, as for the stairs shown in Fig.
84. This may be done as follows:

Measure the curve of the wall, as from **A** to
B, taken at the floor line. This length will
correspond with the natural run, as from **A** to
C. To this, set up the rise of the stairs, which
in this case would be $6 \times 7 = 42$ inches, as
from **A** to **D** in the diagram; and **D** to **B** will
be the required length of the string. The back
of the string can then be kerfed the same as

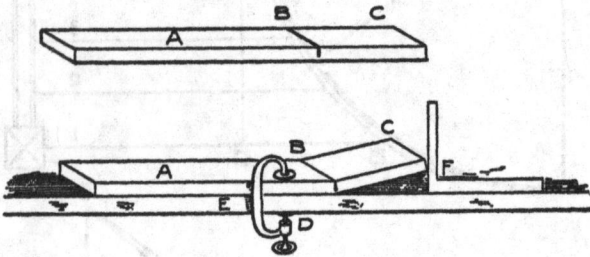

Fig. 58. To Kerf a Circular Member to Any Desired Radius.

for the ordinary base or for a curved riser; but
the kerfs must be cut parallel with the risers.

These diagrams should be laid off full-size,
from which accurate measurement can be taken.

Kerfing a Riser. Saw kerfing is a simple
thing when understood. By the method shown
in Fig. 85, a piece of wood may be bent to any
radius, no matter how thick or thin the material
may be, or how thick or thin the saw may be.

If for a circle to bend, say, three feet in
diameter, take a piece of stuff about one and

one-half inches wide, as **A**, and of the same thickness as the material to be used. Now take the radius of the curve desired—in this case eighteen inches—and make a kerf that distance from the end, as **BC**, to the depth required.

Next clamp **A** down to the bench **E** close to the kerf and raise the radius end till the cut comes together tight, and take the height with the steel square **F** from top of bench to under side of piece. This will give the space between each kerf to bend the riser or any other curved member to the radius desired.

Short Cuts in Stair Work. Short cuts and simple methods of laying out work are always interesting and valuable. Here is a method for laying out strings for a plain, straight flight of stairs.

The sketch, Fig. 86, shows how to get the lengths and cuts—that is, by using the plumb and level at each end of the string, and scribing as at **a** and **b**. Then, by changing the string end for end, it will be found to fit.

First find what rise is wanted; this depends on the space available. If there is sufficient room, steps with 7 or 7½-inch rise are good. Make, for example, the height between the floors 9 feet 2 inches, or 110 inches. Dividing this by 7 gives $15\,^5/_7$ or 15 spaces; and by dividing 110 by 15 will give the exact rise, which is 7⅓ inches.

Since there is plenty of room on the floor for the run, the tread can be made any chosen width. Suppose 9 inches is the desired width. By multiplying 15 by 9, the run on the floor, 11 feet 3 inches, is found, at which point it will

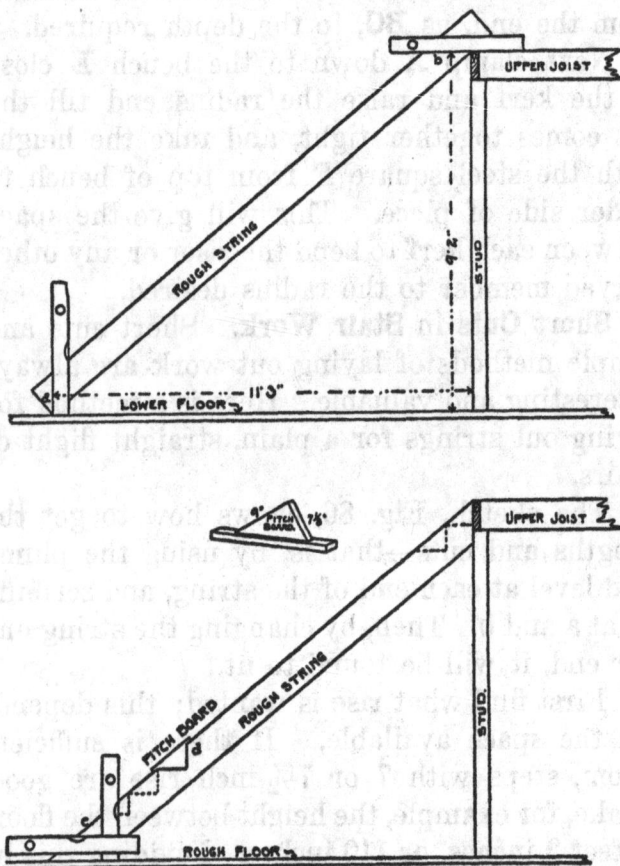

Fig. 86. Short Cut for Laying Out a Straight Stair.

be seen that the lower end of the string has been placed. Then proceed, as already stated, with plumb and level.

After obtaining these cuts, lay the string off into 15 equal spaces with dividers, and again use the plumb and level in laying off the first notch, as shown at **C**. Then carefully cut this out and use the piece for the pitch-board, from which proceed to lay off the rest of the string, as shown at **d**.

Paneled Moulding for Stair Finish. Moulding applied in panels to the stair string makes a very effective finish. In Fig. 87 is shown the face of a stair string with diamond-shaped panels. To begin with, all cuts of mouldings of this kind should be made in a miter box. Then the question is, how to find the angle on the square by which to make the proper cut on the miter box.

In the lower part of the figure is shown an enlarged panel with the application of the square for finding the miter line, or the line of juncture of the moulds. In this, it will be seen that there are two angles in the panel. The sharper one is known as an **acute,** and the other as an **obtuse** angle. The former is less than 90, and the latter more than 90 degrees. However, by this method, finding the miter line for one, also gives the other, because when the blade is giving the miter line for the acute, the tongue is giving it for the obtuse.

In this figure the application for both is shown, and it will be seen that the blade and tongue of like squares are resting parallel with each other. Now, as to the placing of these

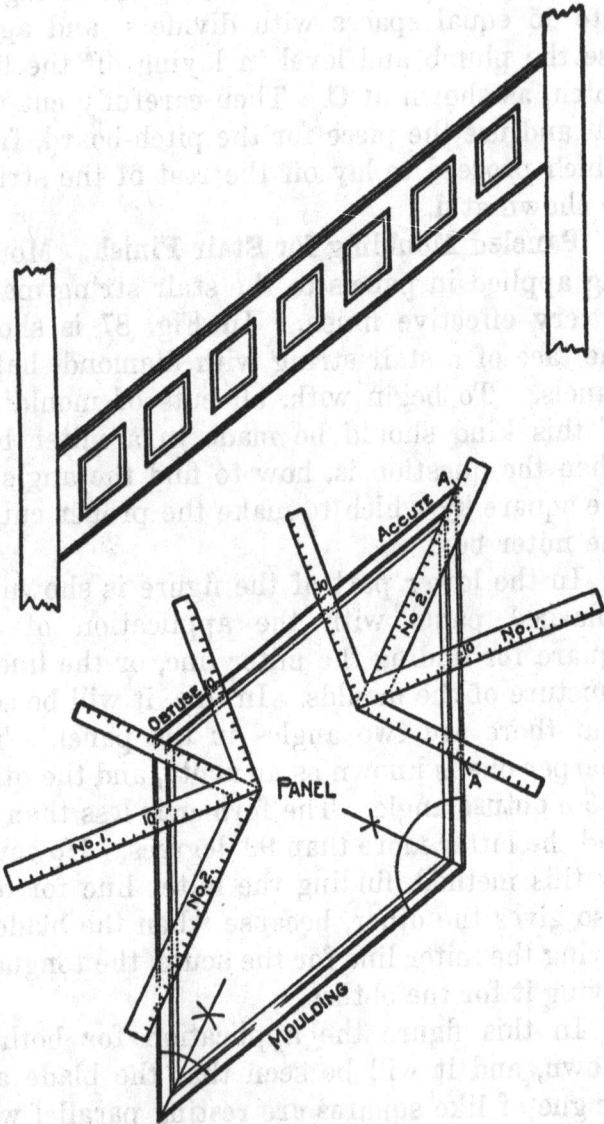

Fig. 87. Paneled Finish for Stairs.

squares to get the proper angle, simply take like
figures on square No. 1, say 10 and 10; place
them so that these figures are at the edge of
the panel, as shown, and with the heel of square
No. 2 resting on the former, and with the blade
intersecting the corner of the panel, the figures
intersected by the edge of the panel, as at **AA**,
will be the figures to use to make the cut on the
miter box for either angle.

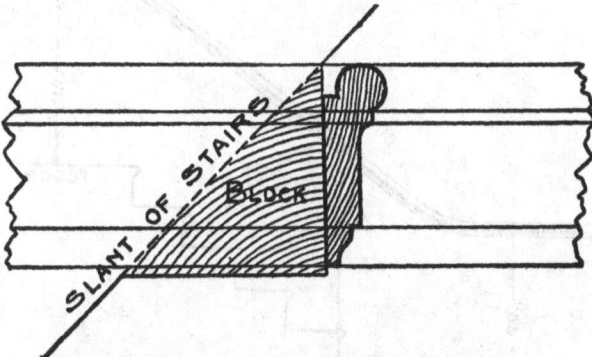

Fig. 88. How to Apply Moulding under Stairs.

The same results may be obtained by bisecting the angle with the compass, as shown in
the two lower corners of the enlarged panel in
the lower part of Fig. 87, and setting a bevel
square to the angles thus formed. This will
give the miter. A more common way, however,
among workmen, is to take a block of any convenient width, and, with parallel edges and with
it set each way in the corner, mark along the
outer sides, the crossing of these lines will be
the point for the bisecting or miter line.

Picture Moulding under Stairs. The right way to put up picture moulding, where there is a stair carriage running up through the room, is a point that sometimes gives trouble, the intersection at the out corner being the cause.

In places of this kind, the moulding pieces will not member, because they are resting at different angles. To have them member, it is

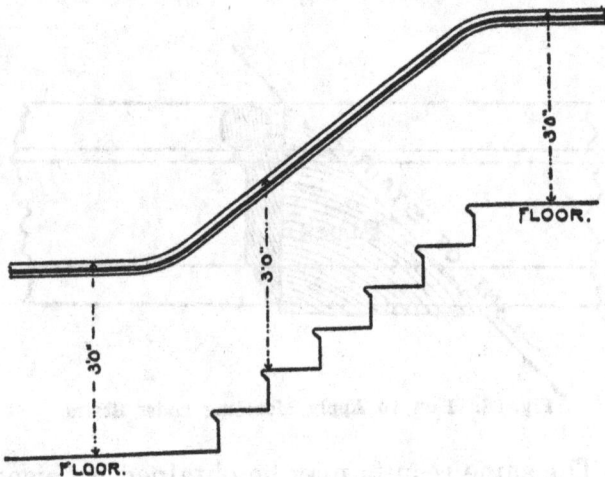

Fig. 89. Height of Wainscoting.

necessary to work a special mould to go under the slanting part of the stairs. Of course, this piece would not be of any use further than to have a continuous mould running around the room. The only way the regular mould can be made to member is to block out as shown in Fig. 88. The back of the mould in that case sets perpendicular, and the angles are the regular 45-degree miter.

Height of Stair Wainscoting. Wainscoting running up the stairs should correspond in height to the wainscoting on the walls. The accepted custom is to continue the same height up the stairway as is wanted on the level part. The measurement should be taken at a point directly above the face of the risers, as shown in Fig. 89.

CEMENT PLASTER HOUSES

Relative Cost and Desirability. Cement siding has grown wonderfully popular in very recent years. The artistic effects which its use makes possible have had much to do with its ready reception by architects, builders, and owners. It furnishes the added advantage of being fireproof.

When properly applied, it is economical in that it will outlast wood or shingle siding and will not require constant painting to keep it from deteriorating. The claim is made, that a good cement exterior will wear better than stone, and will become better both in color and in weather-resisting qualities with age.

The first cost of a cement siding is somewhat in excess of wood siding, painting not considered; but with the ever-growing scarcity of good, clear pine siding, this margin is rapidly diminishing.

Siding determines the life of a house, and the denuding of the pine forests of the United States makes it imperative that a new material

be found to take the place of wood siding. No
one realizes this more than the carpenter, who
has seen the changes which have taken place in
the grading of lumber these last twenty or more
years.

Time was when a C grade of pine siding was
good enough for almost any house; but the
sappy, knotty, blue stuff which passes for that
grade to-day, and the advanced price of this, are
sufficient to make the conscientious builder serious. The tendency of sapwood to push off the
paint, and the readiness with which it decays,
make it questionable whether its cheapness, as
compared with other grades and other materials, is not often overestimated.

The discovery, in almost every part of the
country, of the raw materials from which good
Portland cement is made, and the consequent
rapid growth of mammoth cement mills, are
bound to make cement plaster more available
than ever before. The steady lowering in the
price of Portland cement which has accom-
panied the development of improved processes
in America, has already made its cost very low
as compared with the price formerly paid for
the imported article. It must be admitted that
there has been no small amount of prejudice
aroused against the use of cement siding, because of past failures. It must also be admitted
that the problem of its use is not entirely solved
to-day. It is a fact, however, that its use is
understood well enough, and its success suffi-

ciently demonstrated, to warrant its use on innumerable costly buildings throughout the country. The manner of mixing, the proportion of parts, the coloring, the application and care of the walls after the plaster has been applied, make of it a problem which requires expert skill in handling. An inexperienced workman—unless he gives the matter the utmost careful study, and acquaints himself thoroughly with the methods of approved practice—will be certain to come to grief, causing regret to the

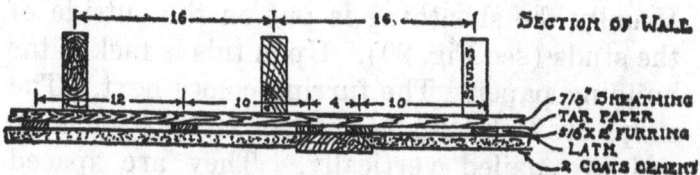

Fig. 90. Wall Section, Plastered House, Half-Timber Effect.

owner, and creating prejudice against cement as a siding material.

The effects which may be obtained are various and interesting. Cement siding may be colored or left natural. It may be finished smooth like the ordinary sand finish of common plaster, or it may be stippled. Rough-cast finish is obtained by throwing pebbles mixed with thin cement upon the wall before it has had time to harden thoroughly. Cement siding may cover the house entirely; or it may be combined with wood, brick, or stone to form the wall. A very popular effect is obtained by using wood siding for the lower, and cement plaster for the upper part of the house.

Artistic effects in English half-timbered houses are due to the ease with which the spaces may be proportioned and arranged. There is an added advantage in the half-timber, in that the material in the smaller spaces is not so likely to check with the expansion and contraction caused by atmospheric influences.

Framing for Cement Plaster Houses. The construction of the frame for a cement exterior differs but slightly from that for wood siding. Usually the sheathing is put on the outside of the studs (see Fig. 90). Upon this is tacked tar building paper. The furring comes next. The strips are ½ by 2¼, ⅝ by 2, or ⅞ by 2½ inches, and are nailed vertically. They are spaced twelve inches from center to center for the lath, irrespective of the position of the studding. The thicker furring is used when more air space is desired than can be obtained with the thinner strips.

That there may be plenty of clinches for the plaster, the wood lath are but one inch wide. They are the usual length, however, and the furring strips are spaced one foot from center to center, just as for metal lath. Fig. 91 gives sectional views of a plain plastered wall and of a half-timbered wall, showing details of construction.

Window and Door Frames. Window framing for cement plaster houses should be set as de-

tailed in Fig. 92, which shows the use of
expanded metal lath secured to strips. This
gives a better clinch for the mortar than if the
lath were stapled directly to the sheathing;
besides, it creates an air space and provides a
wider jamb at the windows, which is essential
where large plate glass is used, necessitating

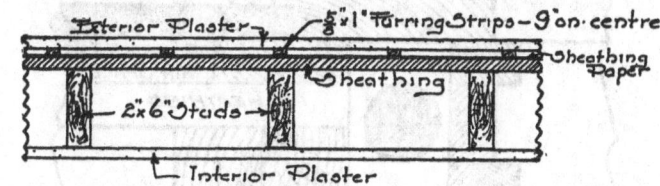

PLAIN·PLASTERED·WALL

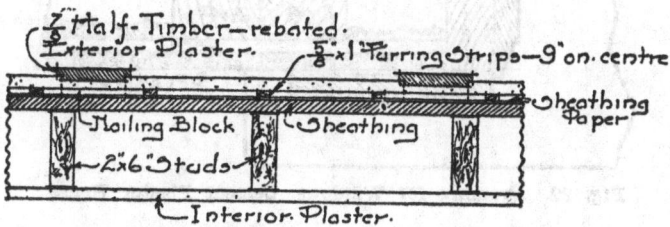

HALF·TIMBERED·WALL

Fig. 91.   Wall Sections—Cement Plaster Houses.

heavier sash than for the common double-
strength glass.   It is a good idea to plow or
groove out the corner of the frame so that the
mortar will extend under the edge of the frame.
The flashing of the caps may be put on in the
usual way, and plastered over.   Of course it
would be much easier, so far as the plastering is

concerned, to set the frames after the plastering is done; but this would not make so tight a job, especially as to the prevention of leakage at the top. The framework should be very substantial; otherwise settlement or vibration will crack the plastering.

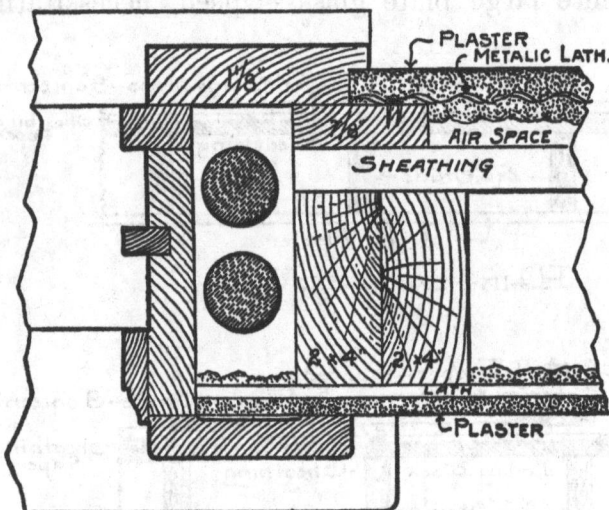

Fig. 92. Framing for Windows, Cement Plaster House.

Casement Window Construction. Fig. 93 shows the construction of an inward-opening casement; also the manner of making and applying stucco to the exterior surfaces of frame walls.

The only serious objection to the use of casement windows in general, is that it is very difficult to make them proof against rain and wind; and with casements opening inward, the difficulty is much greater than with those opening outward. This detail shows as simple a method as can well be employed in constructing an inward-opening casement, though it is not always thoroughly weather-tight during driving rainstorms, when the house is in an exposed location.

The frame is made out of two-inch stock, rebated for the sash, for outside blinds or storm sash. The channel at **B** on both jambs of the frame, is for the purpose of catching any water that may beat in between the sash and the frame, and for conveying it downward to the sill, on which it discharges in the manner indicated by the arrows at **B**. A filling piece **A**, of the same material and finish as the adjoining interior woodwork, is placed on the inner edge of the frame, so that none of the frame will be exposed in the room.

The sill is usually the weak point of inward-opening casements, owing to the fact that whatever rebate is made for the sash, must necessarily have its lower edge on the inside of the window, so that any water which once enters between the lower rail of the sash and the sill will leak into the room. To prevent the water from entering at this point, an undercut drip is provided on the lower rail of the sash, so that any water which may trickle down the outside surface of the sash will drop to the sill from the lower edge; and the sill has a hollow in the raised portion just under the sash, for the pur-

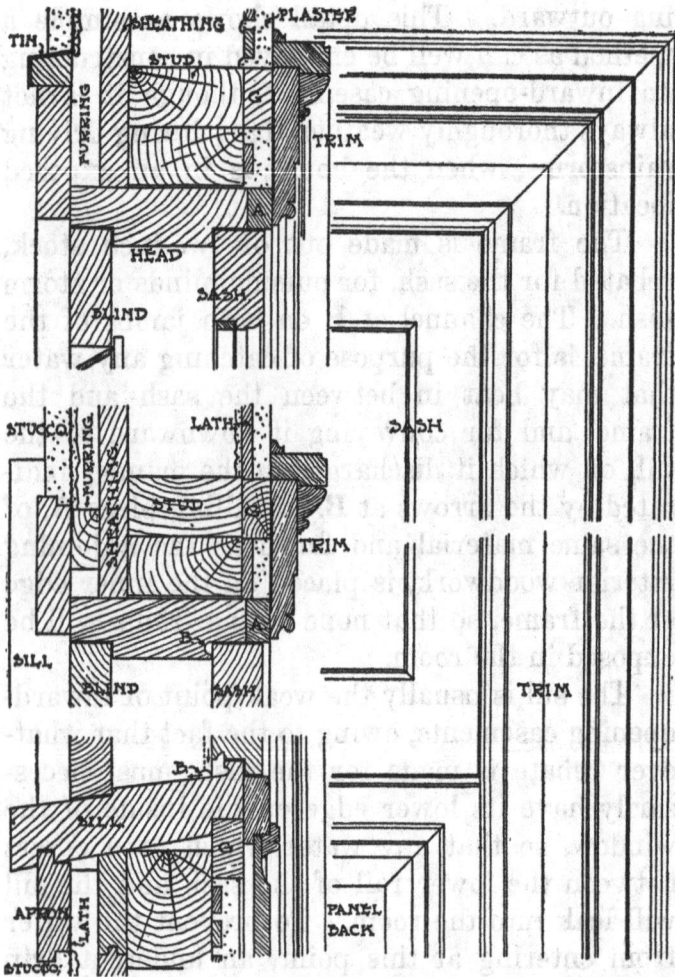

Fig. 93. Casement Window in Cement Plaster House.

pose of casting off any water which may be driven against it in severe weather.

The sill is rebated for the outside blinds, and has an apron which is tongued into it.

The wall is frame, constructed in the usual manner of two by four-inch studs placed sixteen inches on centers, and doubled for jambs, heads, and sills of openings. It is lathed and plastered on the inside; and grounds **G** are set as a nailing for the inside finishing woodwork.

The trim is moulded and hollow-backed and mitered at angles, and has a face-mould. It finishes on wood plinth blocks of the same height as the adjoining base; and underneath the window a moulded panel-back is provided. Walls should be plastered behind panel-backs, but the finish coat may be omitted. Panels should be set loose to allow for expansion and contraction, and the mouldings should be nailed to the stiles and rails of the panel-back. Also, the joint of panels with the stiles and rails should be so constructed that the panel may be readily removed by taking off the moulding. This is very essential, as panels frequently need replacing, owing to cracks and warping.

In order that the wall may be plastered behind the panel-back without increasing its thickness, the studs between the floor and the sill are three by four inch, set flatwise.

The exterior of the wall is sheathed with matched boards surfaced on one side, and is then covered with waterproof building paper, well lapped and tacked. The surface is then furred with one by two-inch strips placed twelve inches on centers and well nailed. Metal lath is then applied, well lapped, stretched, and stapled to each furring strip.

To this lath the first coat of stucco mortar is applied, and it should be well troweled under pressure to secure a good "key" on the lath. A good mortar for this coat is made of one part domestic Portland cement, one part shell lime, and five parts of clean sharp sand. This coat should be allowed to dry slowly, and, if necessary to accomplish this, should be frequently sprinkled with water through a hose having a fine sprinkling nozzle, for the first twenty-four hours. The surface should be lightly scratched, so as to provide a "key" for the second coat.

A good mixture for the second or finish coat is made in the proportions of one part domestic Portland cement, one part shell lime, and five parts of clean white marble dust. This dust should not be the refuse of a marble quarry, but should consist of clean white marble specially ground for stucco work. This coat should be allowed to dry slowly, being dampened if necessary, similarly to the undercoat.

It is essential that the casings, cornice, base, and beltings be so made that the plaster shall be keyed to it. Strips of wood for the English half-timber effect are beveled on their edges as indicated in Figs. 90 and 91. Casings may be similarly beveled on their outer edges except the head, which is tinned so as to turn the water. A more common method of making casings is to run a moulding entirely around the casing, allowing it to project over the outer edge about five-eighths of an inch. Such casings have an "apron" similar to that used on the inside.

Metal and Wood Lath. The question of the relative merits of metal and wood lath is one that does not seem to be fully settled. In fact, both metal and wood have their advantages and their disadvantages. Time will tell. At present both are used in about equal proportion, each having advocates with very decided opinions.

The advantage most frequently urged in behalf of metal lath is its rendering the wall fireproof. Its greatest disadvantage is its liability to rust. This disadvantage, it is claimed by manufacturers, is overcome by having the lath back-plastered so that the meshes are completely embedded. This does not fully protect the metal, however; and to overcome the difficulty, metal lath galvanized or coated with protective paint is being placed on the market.

The advantages and disadvantages of wood lath are too well known to the reader to need repeating. The poor quality of the lath now generally found on the market, which is becoming poorer from year to year, and their liability to shrink, warp, and buckle, render them far from ideal. The decrease in their width, with the consequent more frequent clinches, and their cheapness, have served to keep wood lath to the front in the outlying districts where fireproofing is not so much insisted upon.

Cement Plaster—How to Mix and Apply.

Many manufacturers of cements provide directions for the proper proportioning of their materials. It is taken for granted that their directions are the results of experiments and observations with their products, and they should therefore be considered reliable.

The following, from the annual report for 1904, of the Ohio State Geologist, will be of interest:

First coat, one-half inch thick. For best results, the wall should be furred off with strips put on vertically twelve inches apart and well nailed. On these, fasten firmly metal lath. Add fiber to the mortar for lath work. Wet thoroughly the surface to be plastered. Mix one part of non-staining Portland cement with two parts medium sand, one part fine sand and one-half part lime flour. When this coat has set hard, wet the surface thoroughly, and apply the second coat with a wooden float.

Second coat, one-quarter inch thick. Mix one part cement as above, one part fine sand, and two parts medium sand or crushed granite. Before the second coat has set hard, it may be "joined" to present the appearance of stone work. A small addition of lime flour increases the adhesion of the mortar.

The finished surfaces should be protected for at least two weeks with canvas curtains or bagging saturated with water.

Defects are liable to appear on cement plastered walls, (1) if too much cement is used; (2) if not applied with sufficient moisture; (3) if not troweled sufficiently; (4) if not protected from variations in temperature and draughts of air.

To this a prominent manufacturer of metal lath adds:

In some sections a departure from this specification has been found preferable. It is the practice in the New England States, for instance, to staple metal lath directly to the studding, and then plaster with one heavy coat of Portland cement and lime mortar mixed, using one barrel of best Portland cement and three casks of hair and lime mortar made up in the usual manner, as if it were to be applied to wood lath. The lime mortar to be divided into batches so that the Portland cement can be added in small quantities just before using, that the cement may not have time to harden or set before the plasterer can use it.

After this coat has hardened sufficiently, it is back-plastered on the key formed by the first coat, putting this back-plaster coat on with the same kind of mixture as the first coat on the outside, and covering the lath by at least one-half inch.

After these two coats have hardened sufficiently and dried out, the second or finish coat can be put on, either by slapdashing, or putting on one heavy coat with trowel finish, or applying any of the various attractive finishes which are possible by the use of cement.

The mixture of this final coat depends on the kind of finish desired; but it is usually made with one barrel of Portland cement to two barrels of coarse, sharp sand. If a light color is desired, a hodful of lime putty is added to the mixture; or, if a very rough finish is wanted, a proportion of pebbles or crushed stone is mixed with the sand and cement. It is difficult to give a certain formula for the finishing coat, as nearly every plasterer or architect has his own ideas as to this finish.

The specification last mentioned is the one generally used on the cheaper grades of houses, from $5,000 to $7,500. On the higher grade of work, the plastering is done in four coats. There is first a **scratch coat,** which simply fills the meshes of the lath; second, the **backing-up coat** on the inside; third, what is called the **brown coat,** which is a heavy coat applied directly to the scratch coat, and which is floated or brought up to a straight, smooth surface, and left somewhat roughened to receive the final coat. The brown coat is often omitted on the cheaper class of houses. It is usually mixed with one barrel of Portland cement to two barrels of sand, and a hodful of putty.

It may be added that improper gauging of cement and lime often causes an uneven color. Experienced plasterers overcome this easily. One who has done much of this says he thins down his lime putty so that it is so watery as to be used in mixing the cement.

The accompanying table shows the area which can be covered by one barrel of Portland cement mortar of various mixtures, with coats of various thicknesses.

AREA COVERED BY MORTAR.

Mortar Produced from One Barrel of Portland Cement Mortar (3.8 ou. ft. Cement Paste). (No Lime)

COMPOSITION OF MORTAR	THICKNESS OF COAT	AREA COVERED
1 Cement, 1 Sand	1 inch ¾ " ½ "	67 sq. ft. 90 " " 134 " "
1 Cement, 2 Sand	1 inch ¾ " ½ "	104 sq. ft. 139 " " 208 " "
1 Cement, 3 Sand	1 inch ¾ " ½ "	140 sq. ft. 187 " " 280 " "

Finishes and Tinting. A great variety of finishes is possible. The **stippled effect** is very pleasing; also the effect obtained by throwing small **pebbles** at random into the plaster before the second coat has set. An effective **rough cast** is obtained by mixing cement and water at a thick fluid consistency, and then adding fine washed gravel, screened through a 3/8-inch-mesh screen. When mixed it is ready for application, and may be applied as a third coat on a rough-coated surface, or directly to a scratch coat. The result is most pleasing to the eye, and for a good wearing surface there is none better.

The **color effects** obtained with cement are many and are beautiful. Most of these effects are obtained, however, not as might be supposed, by mixing the dry colors in the cement, but by painting the cement after it has become dry and hard. There are two very good reasons for not mixing the colors in the cement. First, it is almost impossible to mix the mass so that it will dry with an even or uniform color. Second, most coloring matters weaken the cement. No coloring containing acids or anything that will act upon the alkalies in the cement, can be used; and vegetable or oil colors impair the strength of the cement.

The accompanying table indicates the mineral coloring materials which may be used for giving various colors and tints to cement mortar, and the proportions of coloring matter to cement.

MATERIALS USED IN COLORING MORTARS.

Color	Mineral	Pounds Color to 100 Pounds Cement	Pounds Color to Barrel of Cement
Gray	Germantown Lamp Black	¼ to ½	2
Black	Manganese Dioxide	12	48
Black	Excelsior Carbon Black	2	00
Blue	Ultramarine	5 to 6	20
Green	Ultramarine Green	6	24
Red	Iron Oxide	6 to 10	24
Bright Red	Pompeian or English Red	6	24
Sandstone	Red-Purple Oxide of Iron	6	24
Violet	Violet Oxide of Iron	6	24
Brown	Roasted Iron Oxide or Brown Ocher	6	24
Yellow or Buff	Yellow Ocher	6 to 10	24

FRAMING FOR VENEER AND MASONRY HOUSES

Various types of masonry veneer construction have frequently been employed in many parts of the country, especially in the East, and have gained much popularity. Brick veneer has been most used, though of late years concrete blocks and tile have come into use for veneering purposes to some extent. The advantages of any of these systems over the solid masonry wall, for residence work, seems to be a lower first cost and a drier wall; while as compared with all-frame construction, the veneer is warmer, more durable, and presents a more dignified appearance.

Brick Veneer Wall Construction. In the main, the timber framing to be used with brick veneer construction is identical with the best practice for all-frame houses. A number of points, however, require special attention. The relation of the masonry veneer coat to the studding, and the proper bonding of the masonry and framework, one to the other, is one of these points.

Consider a house which is to be constructed of a wooden frame, sheathed diagonally with

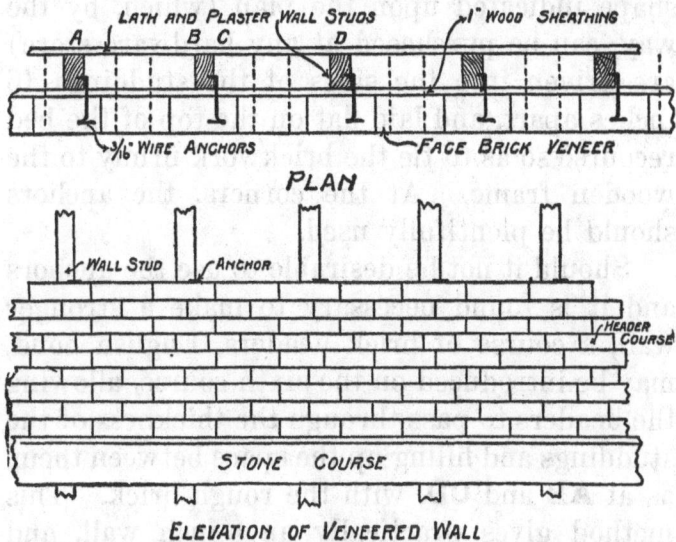

PLAN

ELEVATION OF VENEERED WALL

Fig. 94. Brick Veneer Wall Construction, Showing Bonding.

inch boards and finished with a brick facing of veneer of 4 inches. In order properly to veneer the wooden or timber work, it is necessary that the frame should be kept at least 6 inches from the outside face of the foundation wall. A water-table course of stone should be carried around above the cellar absolutely level, in order

to support the upper structure of brick. Therefore the foundation wall must not be less than
20 inches thick.

The water-table having been set and the
frame erected to the exact measurements, the
first five courses of brick may be laid all the
way around, as shown in the elevation, Fig. 94.
After this is done, wire wall-anchors of the
shape indicated upon the plan (which, by the
way, can be purchased at any hardware store)
are driven into the sides of the studdings 16
inches apart, and laid flat on the top of the bed
recourse so as to tie the brickwork firmly to the
wooden frame. At the corners, the anchors
should be plentifully used.

Should it not be desirable to use the anchors
and it is found necessary to make a stronger
wall, a course of brick headers, English bond,
may be introduced on the sixth course, allowing
the headers to pass through the thickness of the
studdings and filling up the space between them,
as at **AB** and **CD**, with the rough brick. This
method gives practically an 8-inch wall, and
makes a warmer house, as old brickbats can be
used to great advantage. Should a Flemish
bond of headers and stretchers be employed,
then the bricks should be placed as indicated
by the dotted lines shown in the plan. The
thickness of the anchors desired must not be in
excess of the brick mortar joints.

Should the building be of concrete, veneered
with brick, it will be necessary to lay up the

brickwork first, before backing-up of the concrete. All measurements must be carefully watched so that the sills, lintels, bond courses, etc., may be at their proper heights and levels. The same rules apply to backing-up with rough rubble stonework; but it is better to build the stonework first, and, by driving hook anchors into the variegating joints, obtain a fastening in the brick veneer.

Air-Spaces and Bonding. One of the principal boasts made for the brick veneer type of house is concerning its warmth and dryness. This comes from the ample dead air-spaces. A double air-chamber is made by leaving an air-space between the brick and the sheathing.

Fig. 95 shows in sectional detail how this is done. It is a good idea not to crowd the brick close to the sheathing; better set off an inch or the thickness of the blind stop, and make the same wide enough to lap onto the sheathing. However, the building paper should be put on first; and then, after the window-frames are set, it is a good idea to nail a couple of lath an inch or so back of the blind stop, and fill in with mortar, pressing the same in firmly. This can be done at the time the brick are being laid, with practically no loss of time.

The hollow space serves a double purpose, as it affords a dead air-space, and at the same time allows some leeway in correcting unevenness in the framework. It is also a good idea to cut-in pieces between the joist and studding at the different floors, so as to cut off the circulation in case of fire, as well as prevent the movements of the ever pesky mouse.

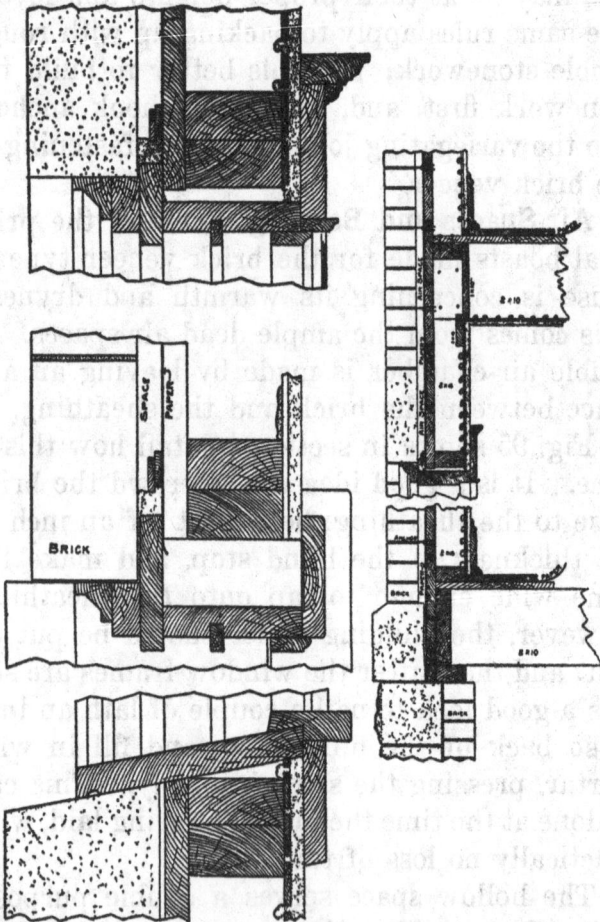

Fig. 95. Wall Section Showing Air Spaces and Framing.

The anchoring of the brickwork to the sheathing should be done by stapling wire to the sheathing opposite the studding and about every sixth course apart. The wire should be left loose enough to reach out half the width of the brick, and to be well bedded into the mortar joint. No. 11 wire should be used.

Window-Framing in Masonry Walls. How window-frames are set, and the woodwork finish attached in masonry walls, is well illustrated in Fig. 96. It shows a casement window opening outward in a thirteen-inch brick wall. This type of construction is about the cheapest that can well be employed, excepting of course that the moulded work and other features which are provided for appearance only may be greatly simplified. These features are subject to considerable modifications as the taste of the architect or builder dictates.

The window opening is spanned on top by a flat stone arch, the blocks of which are cut with a camber of one-quarter of an inch and set with a camber of one-eighth of an inch to every foot of span. Flat arches set in this manner give a much better effect than when set perfectly flat, inasmuch as the arch appears to sag in the center when the soffit is perfectly straight.

Back of the stone arch, a rowlock arch is turned over a wood center, and supports the inner two-thirds of the wall over the opening. These rowlock arches are segmental in form, and are built of brick set on edge; and one rowlock is provided for every foot in the width of the masonry opening. All rowlocks should start

at a brick impost cut to a line corresponding
with the radius of the arch; and the key bricks
of the lower rings should not be set until the
upper rings are ready for their key bricks.

The masonry jamb of the opening is built
straight, and the window-frame is secured in
place by means of a lug which is left on the
jamb of the frame and built into the masonry
as the walls are carried up about same. This
lug also serves as a wind stop.

The stone sill of the opening is cut so as to
lay up accurately with two courses of brickwork, and is tailed into the masonry under each
brick impost. The sill is cut with a wash, and
has a lug or raised seat at each end to receive
the brick imposts. On the under side of the
projecting part, a water drip is cut. The stone
sill should extend under the wooden sill at least
two inches.

At the top, Fig. 96 is a vertical section showing the construction at the head of the frame.
The trim is mitered, put together with slip
tongues, and glued. The head lining is tongued
into a piece of finishing wood on the inside of
the frame head.

Below is a horizontal section showing the
construction at the jamb of the frame. The
frame is moulded, and, where it abuts the stop
bead, a channel is provided to catch any water
which may beat in between the sash and frame
during stormy weather. This channel conveys
the water down, and discharges it on the sill.

The trim is moulded, built up, and hollow-backed, and has a feather-edged back-band.

At the bottom is a vertical section showing

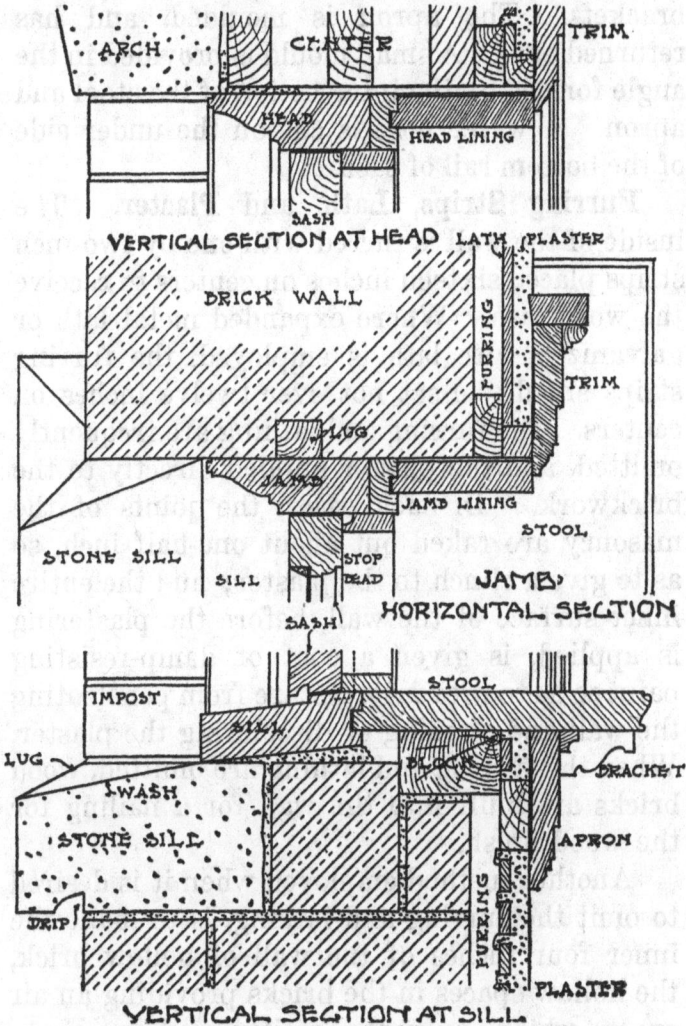

Fig. 96. Casement Window in 13-Inch Brick Wall.

the construction at the sill of the frame. The inside stool is tongued into the wooden sill, extended into the room, and provided with brackets. The apron is moulded, and has returned ends. A small mould is provided in the angle formed by the intersection of the stool and apron. A water nose is cut on the under side of the bottom rail of sash.

Furring Strips, Lath, and Plaster. The inside of the wall is furred with one by two-inch strips placed sixteen inches on centers to receive the wood lath. Where expanded metal lath or galvanized wire lath is employed, the furring strips should be set not over twelve inches on centers. The furring and lathing are frequently omitted, and the plaster applied directly to the brickwork. In such cases, the joints of the masonry are raked out about one-half inch, so as to give a clinch to the plaster; and the entire inner surface of the wall, before the plastering is applied, is given a coat of damp-resisting paint so as to prevent moisture from penetrating the wall and staining or discoloring the plaster. When the furring and lathing are omitted, wood bricks are built into the wall for a nailing for the wood finish.

Another method employed when it is desired to omit the furring and lathing, is to make the inner four inches of the wall of hollow brick, the hollow spaces in the bricks providing an air space which prevents moisture from penetrating.

Framing for Fine Casement Windows. Fig.
97 shows a somewhat better form of construction than that just described, for casement windows opening outward in brick walls. It has
been designed with a view to show only a narrow margin of wood frame about the sashes;
and for this purpose the masonry opening is
rebated to receive the bulky part of the frame.

The wall, which is sixteen inches in thickness, is of brick, and the opening is spanned on
the top with a flat arch of brick ground to the
proper radius. Back of this face arch, a rowlock relieving arch is turned over a timber back
lintel; and the space between the top of the
lintel and the soffit of the relieving arch is filled
in with brick, and is known as the **core** of the
arch. When the head of a window extends
nearly to the ceiling of the room, it is necessary
to provide a steel or cast-iron lintel, instead of
the timber lintel and relieving arch, to support
the ends of the beams.

A stone sill is provided at the bottom of the
opening, and is cut with a wash so as to pitch
off water, and at each end has raised stools or
lugs to receive the brick imposts. It is tailed
into the masonry four inches at each end,
extends under the wooden sill at least two
inches, and has a drip cut on the under side of
the projecting portion.

The window-frame is made from three-inch
by three-inch stock, rebated for sash, and
plowed for jamb and head linings. It is set

in the masonry rebate, and is anchored in place
by means of galvanized wrought-iron anchors
of the form shown, screwed into the frame and
built into the joints of brickwork.

The wall is furred on the inside with one-
inch by two-inch strips set sixteen inches on
centers; and to these strips are nailed grounds
for base, trim, etc., and wood lath for plastering. When expanded metal or wire lath is used,
the furring strips should be set twelve inches
on centers.

The trim is moulded and worked out of one-
inch by four-inch stock, and has a plain back-
band, a face-mould, and a small wall-mould.
Wall-moulds should be small enough to be pliable so as to fit the unevenness of the finished
plaster. The trim finishes on a moulded stool,
tongued into the wooden sill and finished with
an apron. Where deep jambs occur on the
inside of windows, paneled head and jamb
linings are preferable to the plain linings shown.

**Use of Temporary Frames during Construc-
tion.** In important work it frequently happens
that the window-frames are not built in as the
walls are carried up, because of the danger of
damaging them by hauling in materials through
the windows. In such cases, rough timber bucks
about two inches by four inches are built into
the masonry, and the frames are set and nailed
to the bucks after all the rough structural work
of the building is completed. The heads and
sills of the bucks are allowed to project three

VERTICAL SECTION AT HEAD

HORIZONTAL SECTION — JAMB

VERTICAL SECTION AT SILL

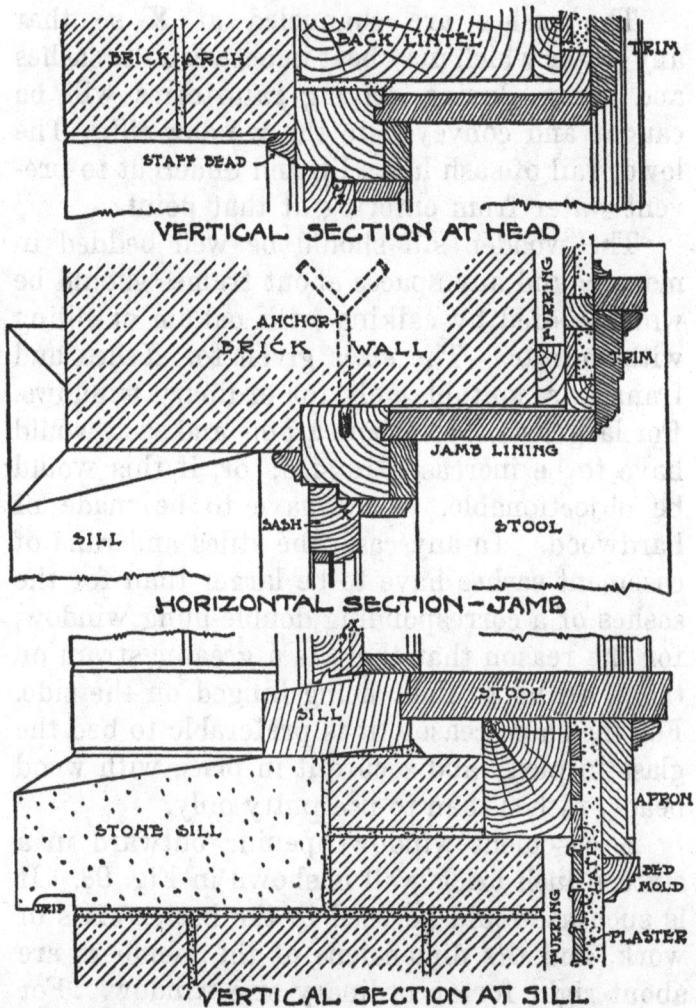

Fig. 97. Casement Window in 16-Inch Brick Wall.

or four inches over the jamb bucks; and the
projecting portions are built into the masonry
so as to securely anchor them in place.

The sashes are channeled at **X** so that
any water which may beat in between the sashes
and frame during driving rainstorms will be
caught and conveyed to the window-sill. The
lower rail of sash has a lip and undercut to prevent water from entering at that point.

The wooden sill should be well bedded in
mortar; and any spaces about frames should be
wind-proofed by calking with oakum or filling
with mortar. The sizes given for sashes and
frames are suitable only for ordinary windows.
For larger windows, both frame and sash would
have to be increased in size; or, if this would
be objectionable, would have to be made of
hardwood. In any case, the stiles and rails of
casement sashes have to be larger than for the
sashes of a corresponding double-hung window,
for the reason that there is a greater strain on
them, owing to their being hinged on the side.
For the same reason, it is preferable to bed the
glass in putty and secure it in place with wood
beads, rather than to use putty only.

A casement window opening outward in a
sixteen-inch brick wall is shown in Fig. 98. It
is such as would be used in the better class of
work, and the dimensions of the members are
about right for an ordinary size window. For
larger windows, the frame would have to be
increased in size, and the sashes made of cherry
or other suitable hardwood, rather than the
dimensions of stiles and rails increased, which
would be objectionable in that it would show too

much wood. The thickness of sashes, however, should be increased. In any case, the dimensions of stiles and rails for casement sashes are greater than for double-hung sashes in windows of similar size, owing to the greater strain on them caused by their being hinged on one side.

The frame shown is built as the masonry walls are carried up, in a rebate formed in the wall, so that too much of the frame will not be exposed to view; and is secured firmly in place by means of lugs housed into the jamb and built into the brickwork. All spaces between the frame and the brickwork are calked so as to be wind-proof, the calking consisting of oakum well compacted and plastered over, or scratch-coat mortar slushed in when the plastering work is being proceeded with. Calking should never be omitted in important work. Frames, after being built in, should be well protected with boarding so as to prevent them from being damaged in passing materials through the openings.

The interior treatment of the window opening is, of course, subject to innumerable modifications, as the taste of the architect or owner dictates and purse affords. The architrave shown is quite effective, and not so costly as the finished effect implies. The trim is moulded and worked out of seven-eighths-inch material, and is blocked at the back to make it heavier in appearance. A moulded back-band adds to its massiveness; and a small flexible wall-mould

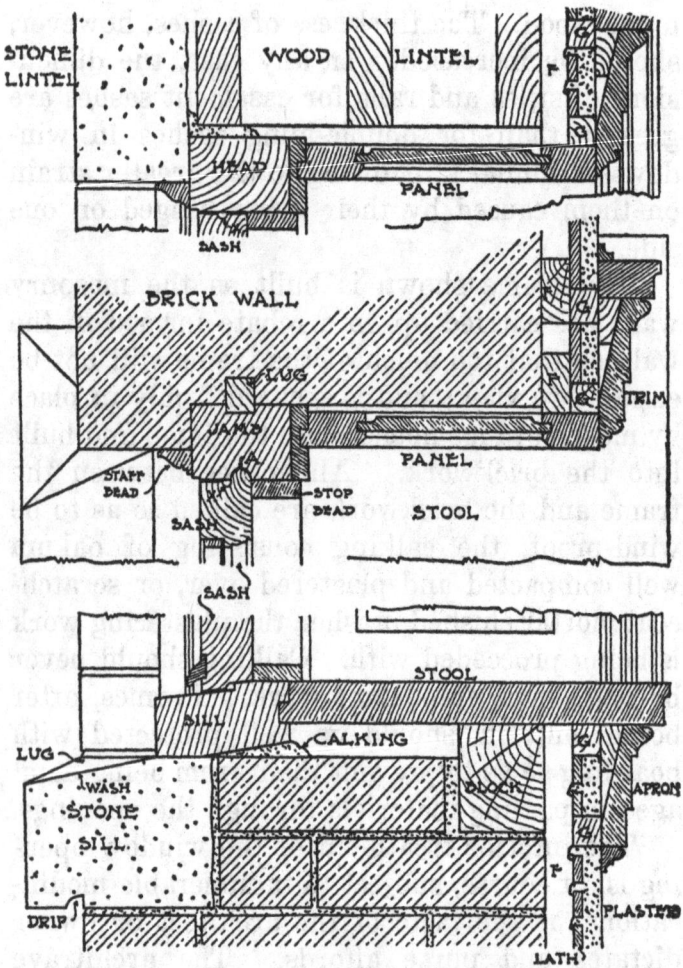

Fig. 98. Outward-Opening Casement in Brick Wall.

covers the junction with the plaster work, and is easily bent to follow the unevenness of the finished plaster.

The exterior of the window consists of brick

imposts showing a three-inch reveal, a stone lintel spanning the top of the opening, and a stone sill across the bottom. The sill has a bed of five inches, a projection of two inches, a thickness equal to two courses of the face brickwork, and a length sufficient to tail four inches into the masonry at each end. The upper surface has a wash and stools at either end, and the projecting portion has a drip cut on the under side.

The inside of the wall is furred with one-by-two-inch strips, indicated by **F**, to which grounds **G** and lath are applied. The head of the window has a paneled head lining tongued into the finishing woodwork which is provided to cover the rough frame. At **X**, on the top rail of the sash, a channel or gutter is provided to catch any rainwater which may beat in between the sash and the head of the frame. This channel is continuous across the head, and conveys the water to a similar groove on the sash stile.

Inward-Opening Casement Windows. Fig. 99 shows a very successful method of constructing an inward-opening casement so that it will be proof against wind and rain.

The jamb of the frame is set in a rebate in the masonry wall, and has a semicircular groove cut in its outer edge for a corresponding semicircular tongue on the stile of the sash. The sash tongue fits snugly into this groove, and makes a perfectly weather-tight joint. This

form of construction requires that the hinges
or butts shall be set so that their pins are from
one-quarter to three-eighths of an inch inside

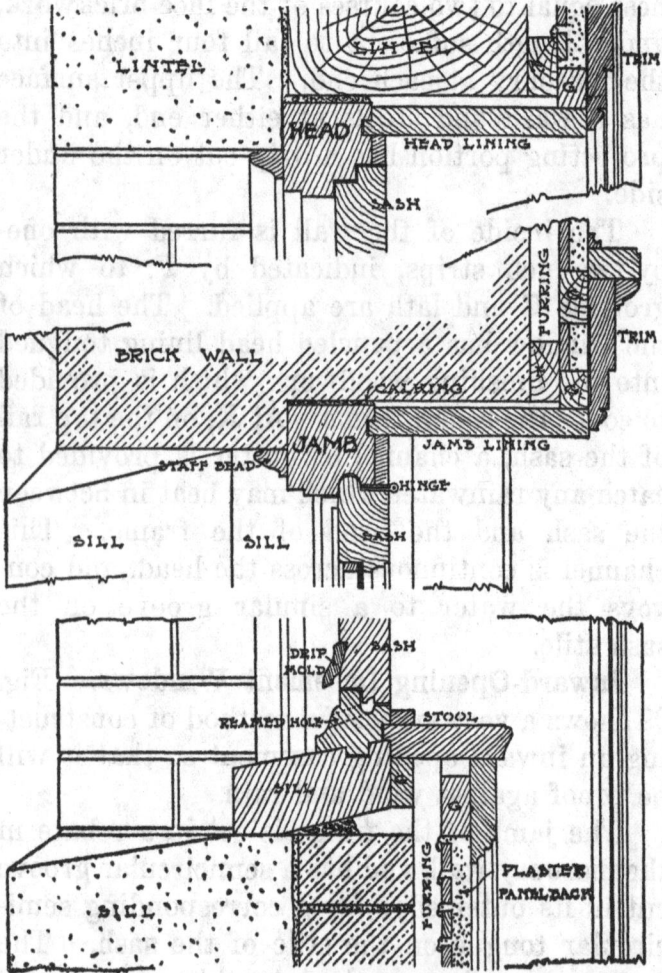

Fig. 99. Storm-Proof Inward-Opening Casement.

of the inner surface of the sash, so that when the sash is opened it will turn on a center sufficiently away from the sash to throw the sash slightly into the room and prevent binding at the tongue and groove.

The head of the frame has a double rebate, and the top rail of sash a single rebate, to exclude the weather.

The joint of the sill and the bottom rail of the sash in inward-opening casements is a particularly difficult one to make weather-tight, and we know of no better way of constructing it than that shown in the illustration. Windows constructed in this manner have remained tight through driving rainstorms.

A moulded member is placed over and tongued into the top inner edge of the wood sill, and is rebated for the bottom rail of the sash. This member for its entire length has a semi-circular groove or gutter in the rebate, as shown, to catch any water which may beat in at the junction of the sash and sill. At intervals of about one foot, reamed holes are provided from the gutter to the outer surface of this strip, as indicated by the dotted lines in the illustration, Fig. 99, to carry away and discharge on the sill any water which may accumulate in the gutter. The holes should be reamed perfectly smooth, and painted; and the joint of the sill and the member immediately over it should be made in white lead.

The bottom rail of sash is rebated, and on

its outer face has a drip-mould let in and joined
in white lead. This mould, under ordinary conditions, will prevent water from entering under
the sash. In driving rainstorms it may not
prevent a little water from working in; but any
such water will be caught at the undercut drip
on the bottom surface of sash, and will drop
into the gutter in top member of sill.

A small moulded staff bead covers the joint
of the masonry and the window-frame; and all
interstices about the frame are calked with
oakum, as indicated, so as to be wind-proof.
Jamb and head linings are tongued into the
frame, and, where deep inside jambs occur, are
better paneled.

The trim is worked out of seven-eighths inch
material, moulded and hollow-backed, and provided with face-mould, back-band, and small
flexible wall-mould. Joints at angles are mitered
and put together with slip tongues, glued, and
screwed. In the better class of work, the trim is
put together at the carpenter shop; and all surfaces, including backs, edges, ends, splines, and
faces, are primed. It may then be brought to the
building without risk of damage through the
dampness in the atmosphere or at the building.

The trim is carried to the floor, and finishes
on moulded wood plinths. The inside recess of
the window is carried to the floor, the masonry
wall being made four inches thinner under the
window. A stool and apron are provided, but-
ting into the jamb linings; and the base breaks

around the recess, thus forming a plaster panel-
back. This plaster panel-back may be painted
or grained to match the adjoining woodwork,
and the effect of a wood panel secured.

The opening in the masonry wall has a stone
sill and lintel, and the inside of the wall is furred
with one-inch by two-inch strips. Grounds G
are set wherever required for a nailing for the
interior finishing woodwork.

Framing for Casement, Screen, and Blind.
Fig. 100 shows the construction of an inward-
opening casement window in a brick wall, with
insect screens placed outside of the sashes, and
with blinds placed outside of the insect screens.
It is also arranged so that when the screens are
removed in winter, storm sashes may be installed
in their place.

The frame is set in a very slight rebate in the
masonry wall, and is secured in place by means
of the lug on the jamb of the frame. This lug
is built in as the brickwork is carried up.

The window-frame is rebated for the sash,
and is also rebated for a tongue on the edge of
the sash. This tongue is quarter-round, to allow
for the play of the sash when opened. The
jamb lining is tongued into the frame, and should
be placed sufficiently back from the jamb of the
frame to allow for the window-shades, which, in
the case of inward-opening casements, are placed
on the top rail of the sash.

The top edge of the sash is slightly beveled

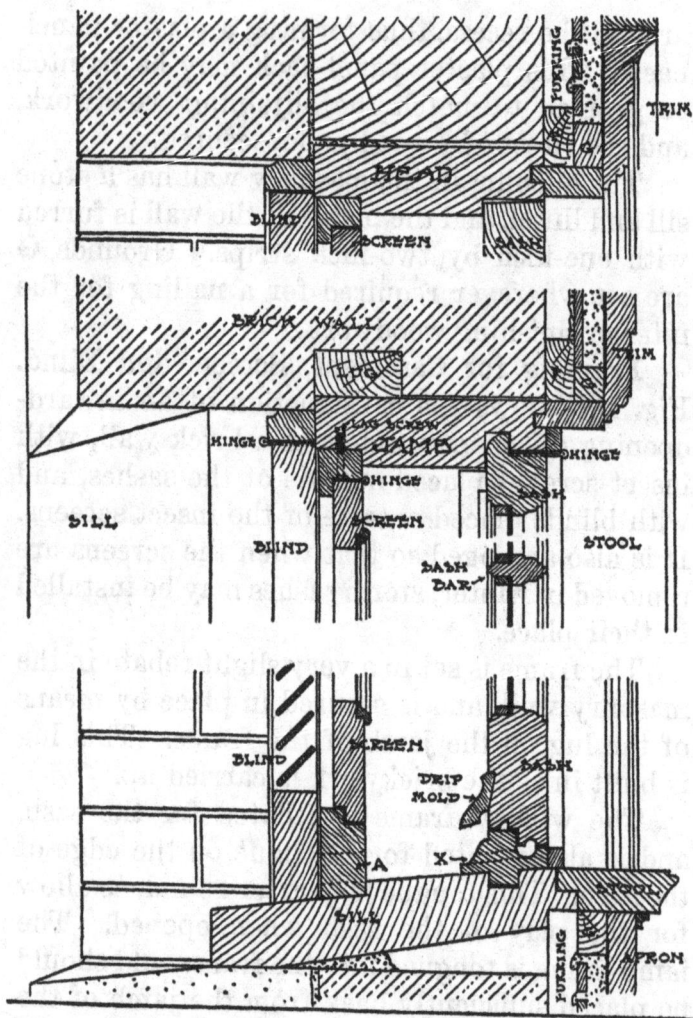

Fig. 100. Casement, Screen, and Blind in Brick Wall.

so that there will be no chance of it striking the transom bar when being closed.

The sill construction is similar to those shown

previously, with undercuts and drip-mould to prevent the entrance of rain water at the joint of the sash and sill. This joint is usually the weak point of inward-opening casements, but, if constructed in the manner shown, will resist driving rainstorms. The inside stool is rebated over the sill.

The transom sash, which is hinged at the bottom and swings in at the top, is rebated over the transom bar and provided with a drip-mould similar to the lower sashes. The top edge of the sash is slightly beveled to allow for the upward throw of the sash when being opened. The transom bar is moulded, and has an undercurrent just below the transom sash to cast off any water which may be driven against the joint (see Fig. 101).

The meeting rails are rebated and beveled, and have inner and outer astragals or cover-moulds. At **X** the sash is grooved to catch any water which may work partly in at the joint.

The window-frame is rebated on the outer edge for the mosquito screen frame, which is secured in place by means of brass lag screws. These screws are so arranged that they may be used to secure storm sashes in place when the screens are removed in winter. When the storm sashes are in place, the blinds cannot be operated from the inside; so they must remain open, unless the blinds are fitted with a device to open them from the inside of the house.

There are several of these opening devices

on the market, and they consist of a worm-gear apparatus which opens the blinds by the turning of a crank within the room, and without the necessity of opening the sashes. These devices are thoroughly practical and very useful, even

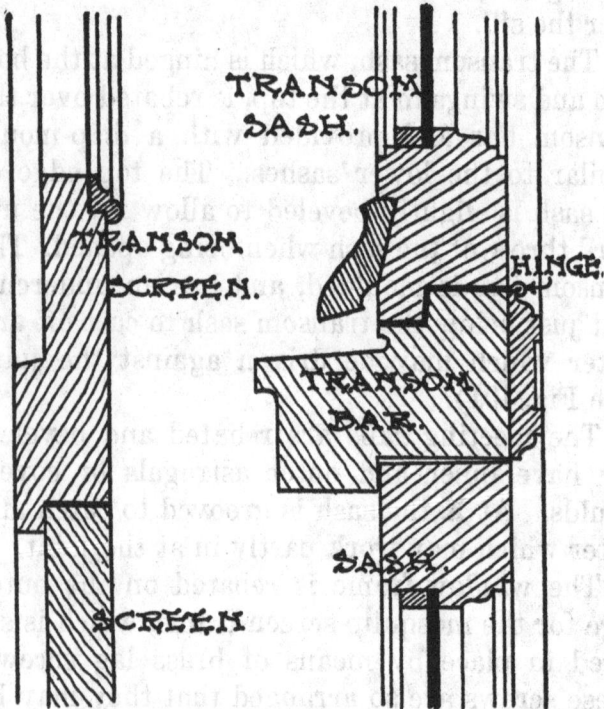

Fig. 101. Section at Transom.

in cases where they are not absolutely necessary. They permit of opening and closing the blinds in stormy weather without opening the window and subjecting the person and the room to the storm.

The screens are rebated, and hinged at the side to open in. The opening in the frame at **A**, bottom of Fig. 100, is covered with netting and allows any water which may come through the screen to pass out over the sill. The transom screen is stationary.

Figs. 100, 101, and 102 respectively show sections through the head, jamb, and sill; transom bar; and meeting rails.

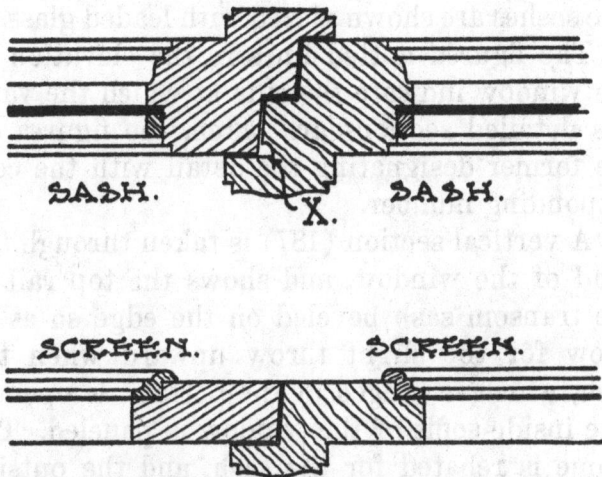

Fig. 102. Section at Meeting Rails.

Casement Bay Window Construction. Fig. 103 shows a bay window in a stone wall, with sashes of the inward-opening casement type. The bay window is entirely within the thickness of the wall, as shown in the plan.

The wall is constructed of random-coursed, roughly squared local stone, with cut quarry

stone sill and lintel. The sill is cut with a wash
and with stools at either end, is tailed into the
masonry at each end, and extends to within two
inches of the inner face of the stone wall. The
masonry jambs of the opening are straight, and
the frame is secured in place by means of lugs
left on the ends of both head and sill, and built
into the stonework as the walls are carried up.
This requires very careful calking of all crevices
about the frame, so as to make a wind-proof job.
The sashes are shown glazed with leaded glass.

The figured dotted lines in the elevation of
the window indicate the cuts at which the various detailed sections are taken, the figures on
the former designating the detail with the corresponding number.

A vertical section (187) is taken through the
head of the window, and shows the top rail of
the transom sash beveled on the edge so as to
allow for the slight throw upward when the
sash, which is hinged at the bottom, is opened.
The inside soffit of the window is paneled. The
frame is rebated for the sash, and the outside
casing is moulded as shown.

A vertical section taken through the transom
bar of the window is given at 188, and shows
the transom sash and bar with a rebated joint,
and the sash with an undercut. The transom
bar is also rebated for the casement sash, and
has an undercut on the projecting portion. A
moulded member of finishing woodwork covers
the transom bar on the room side.

fer so great a resistance to storms and cold as do the horizontally pivoted casements.

The frame is cut out of $2\frac{1}{4}$-inch stock, moulded, and tongued for inside head and jamb linings. The masonry opening is constructed with straight jambs, and the frame is secured in place by means of lugs on the jamb of the frame, which are built into the masonry as the walls are carried up. The section through the jamb is similar to the section through the head of the window. The lug which is indicated there by the dotted lines occurs only on the jambs and not on the head, and is only shown in the top part of the figure to indicate wherein the head and jamb sections differ.

This section at the top (Fig. 104) shows the head lining tongued into the rough frame, and a cover-mould in the angle of head lining and frame. The furring on the inside of the wall is of 1 by 2-inch strips placed 16 inches on centers for the wood lath, or 12 inches on centers for expanded metal or galvanized wire lath. Grounds G are set wherever required for a nailing for the interior wood finish or as a gauge for plastering. The trim is moulded and hollow-backed, and has a back-band and a small wall-mould. This wall-mould follows across top of base and across top and bottom of chair rails where such occur.

The masonry opening is spanned on the exterior by a stone lintel, and back of this a timber lintel. A brick relieving rowlock arch is turned over the timber lintel, one rowlock being pro-

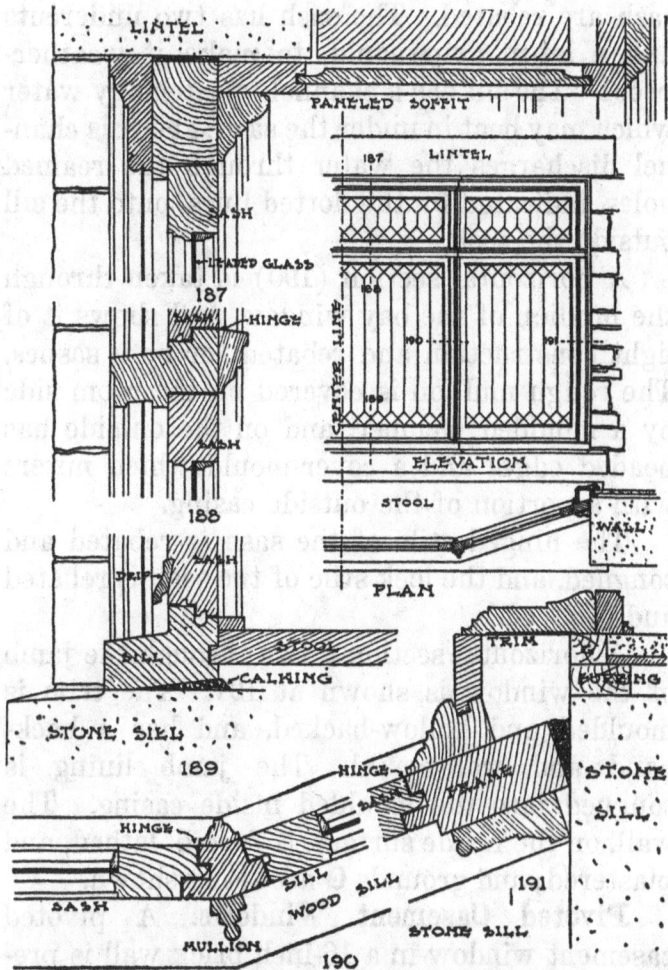

Fig 103. Inward-Opening Casement Bay.

A vertical section (189) is taken through the sill of the window, and shows a method of construction similar to those already illustrated. The stool is tongued into the sill, and the sill and

In the middle portion of Fig. 104 is illustrated a vertical section taken through the window at the axis of the sash; it shows the window closed, by the solid lines; and open, by the broken or dotted lines. The outside and inside stop beads, marked **C** and **D**, are cut at an angle of 45 degrees at **A** and **B**; and half of each stop bead is fastened on the frame, and the other half on the sash, as indicated by the dotted lines which show the sash open.

The projecting part of the jamb of the frame between the two stop beads **X** is cut away between the horizontal dotted lines, shown a little above **A** and a little below **B**, to allow the sash to turn.

At the bottom in Fig. 104 is shown a vertical section taken through the sill of the window; it shows the joint of sash and sill rebated. The stone sill is cut with a wash, has stools at either end, and extends under the wood sill two inches. The inside is finished with a stool and a moulded panel-back. The furring, lathing, and plastering are carried in back of the panel. The trim extends to the floor, finishing on moulded stools.

A pivoted window in a sixteen-inch brick wall is illustrated in Fig. 105. The sash is center-pivoted at top and bottom, and set in a rebated frame two and a-quarter inches thick.

The masonry opening is spanned on top with a flat stone arch, the key of which projects beyond the face of the wall. Back of this arch, a steel lintel consisting of two 3 by 4-inch angles is provided to support the masonry. The inside of the wall is furred with two-inch ribbed full porous terra-cotta blocks, to which the plastering is applied. Grounds **G** for the wood finish are nailed to this furring, which, being full porous, readily receives and holds a driven wire or cut steel nail.

The joint of the wood frame and the masonry is covered with a moulded staff bead. The inside head and jambs are lined with seven-eighths inch material tongued into the frame.

Small wood moulds cover the joints between the sash and the frame, both on the outside and the inside, so as to make it weather-tight, forming a rebate, as shown. As part of the sash opens outward, and the other half into the room, these moulds are fastened to the frame in some places, and to the sash in other places.

At the head, the mould on the outside of the window has half its length fastened to the left side of the sash, as at the dotted lines **A**; and the other half is fastened to the frame. With the inside mould at the top of the window, the reverse is the case, the mould having half its length fastened to the right side of the sash. This is also the case with the inside mould at the bottom of the window, except that this mould is cut as at **B**, and then slit horizontally as indicated by the dotted lines at **C**.

The projecting member of the frame at **D** is cut away on the dotted line for the distance indicated by **F**, so that the ends of the mouldings

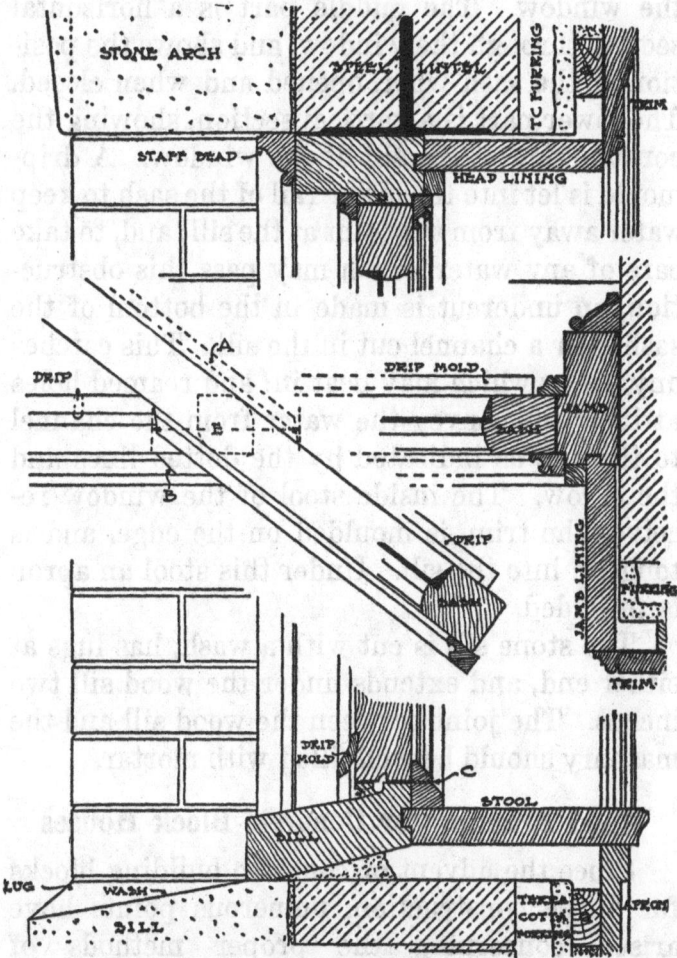

Fig. 105. Pivoted Casement Window.

which are fastened on the head of the sash and
project above it will clear the frame at this point.

In Fig. 105, the upper portion is a vertical
section, showing the construction at the head of

the window. The middle part is a horizontal section through the window, and shows the position of the sash when opened and when closed. The lower part is a vertical section, showing the construction at the sill of the window. A drip-mould is let into the lower rail of the sash to keep water away from the joint at the sill; and, to take care of any water which may pass this obstruction, an undercut is made in the bottom of the sash over a channel cut in the sill. This catches any water which may beat in; and reamed holes at intervals convey the water from the channel to the sill, as indicated by the dotted lines and the arrow. The inside stool of the window receives the trim, is moulded on the edge, and is tongued into the sill. Under this stool an apron is provided.

The stone sill is cut with a wash, has lugs at either end, and extends under the wood sill two inches. The joint between the wood sill and the masonry should be well filled with mortar.

Wood Framing for Concrete-Block Houses

Since the advent of concrete building blocks for house construction, numerous points have arisen concerning the proper methods of framing to be used with them. How to arrange the window-frames with pockets for weights for a dwelling house built with an eight-inch wall, is one of these points. Also, what is the best way to attach a hip-roofed porch to a building of this kind.

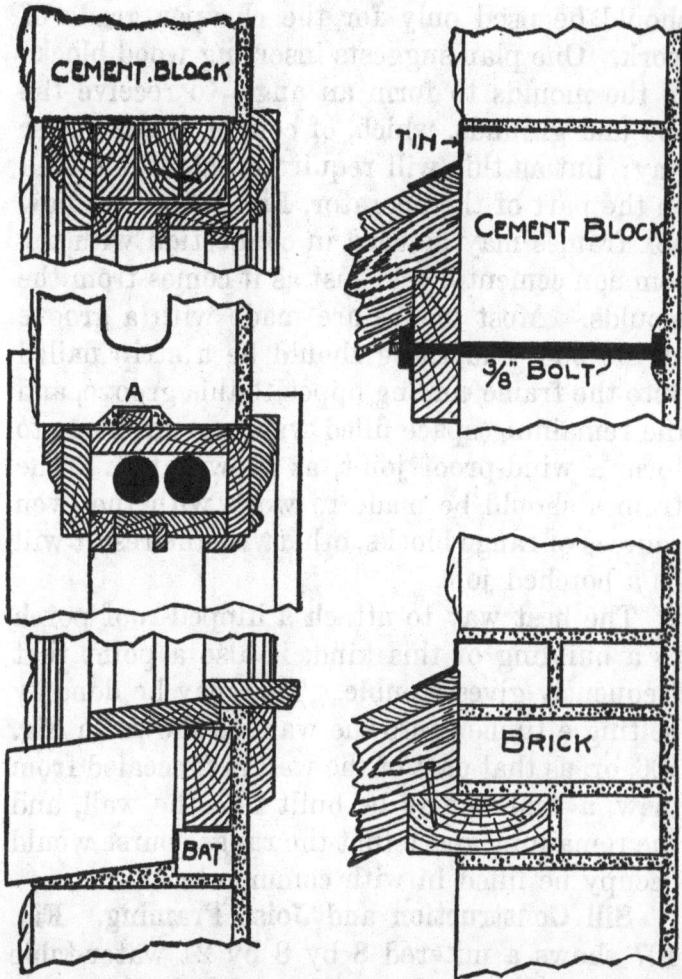

Fig. 106. How to Arrange Window-Frames and Attach Porch Roof.

The thickness of the wall (8 inches) necessarily crowds the frame in giving the proper space for the box to contain the weights. In fact, it is too close for the best class of work, and

should be used only for the cheaper grade of
work. One plan suggests inserting wood blocks
in the moulds to form an angle to receive the
box and grounds, which, of course, is the better
way; but as this will require considerable skill
on the part of the operator, Fig. 106 shows how
box frames may be used in connection with the
common cement block, just as it comes from the
moulds. Most blocks are made with a groove
at the end, and there should be a strip nailed
onto the frame coming opposite this groove, and
the remaining space filled with mortar, so as to
form a wind-proof joint, as shown at **A**. The
frames should be made to work with the even
courses of range blocks, otherwise the result will
be a botched job.

The best way to attach a hipped-roof porch
to a building of this kind, is also a point that
frequently gives trouble. This may be done by
bolting a timber onto the wall, as shown in Fig.
106, or, as that part of the wall is concealed from
view, a timber may be built into the wall, and
the remaining space that the range course would
occupy be filled in with common brick.

Sill Construction and Joist Framing. Fig.
107 shows a mitered 8 by 8 by 24 water-table
of a concrete-block residence. **B** is the same,
except it is 6 by 8 by 24; **C** is a common 8 by 8
by 24 water-table; and **D** a 6 by 8 by 24; **E** is
a 4 by 8 by 14 filler between joists; **F** is a 1 by
7 string-board nailed to end of floor-joists, which
acts as a spacer to hold joists in position till

filler blocks are laid, and serves to aid one in getting the tops of the joists exactly level, as the concrete blocks are not .always perfectly true. If joists are laid on blocks, their imperfections cause some places to be lower than others, and you are in trouble, as the joists cannot be raised after the blocks are laid above them. **G** is the floor-joist; and **H** is a piece of 2 by 8 gained down level with top of floor-joist, to be used at a door opening onto a porch; however, on a porch the water-table course should be left off, and plain-faced blocks used instead. The dotted line **L** is the line of wall under water-table, the water-table projecting two inches over the wall. **JJ** is the double header to support floor-joists at the basement windows. **K** is an 8 by 8 concrete lintel over basement window.

This figure also shows the style of concrete block used around windows, together with a box window-frame, and the way it is set to the concrete wall, and the way the inside casing is fastened to the frame and also to the concrete wall. There is also a wooden wedge driven in the mortar joint next to the frame; the casing is nailed to this.

The window-sills are 8 by 8 concrete and project 2 inches outside of the wall. This leaves a space 2 inches deep under the window, which may be filled with a 2 by 8 with the ends slightly beveled, so as to fit tightly between the blocks on either side of window; this makes a good place to nail the apron to.

One of the worst problems connected with concrete block work is to fasten the head casing to the concrete lintels over the windows and doors. A good method is as follows: In making the lintels, prepare some ¾-inch round pins, 2½ inches long, and soak them in water for at least forty-eight hours. Be sure they are of the softest wood obtainable, and non-resinous, so that they will swell up freely. Now bed these in the lintels while making. They will shrink and drop out by the time the lintel is

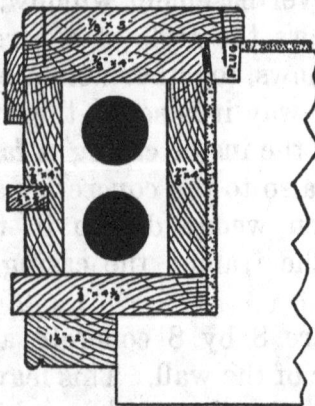

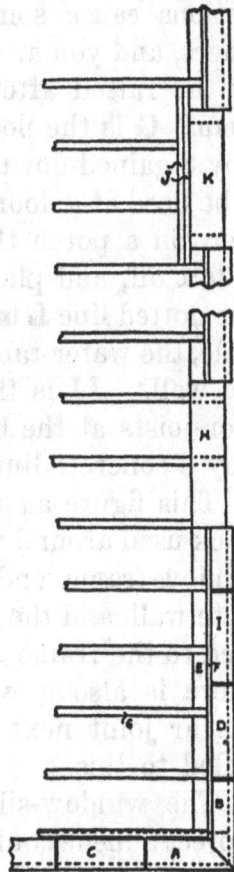

Fig. 107. Wood Framing with Concrete Blocks.

cured enough to put in the house. Place about two to each lintel at the ends of the head casing to be nailed; and by the time you are ready to case the inside, you can drive soft, dry wooden plugs in the holes, and they will hold perfectly well. The reason for wanting the plugs soaked well, is because otherwise they will swell and break the lintel while it is still green.

Attaching Woodwork to Concrete. The best method of securing furring to brick or cement-block walls has caused not a little discussion. Some contractors prefer plucks instead of joint strips, as the strips expand when built in the wall, and afterward become loose on account of shrinkage. A good way is to use a heavily barbed nail driven into the mortar joints. This proves durable, and requires no previous preparation.

The use of small hardwood (well-seasoned) wedges driven into the mortar joints, also serves for attaching casing. Many builders no longer fur on concrete blocks, but apply plaster directly; this saves plaster, and, with proper precautions as to waterproofing to prevent water soaking through or condensing on the inside, will make a good job.

Tile Veneer for Frame Buildings

As modern buildings supersede the old-style frame buildings, it often becomes a question how to remodel them. An effort to imitate the new is seldom successful; but if the old buildings are improved in their own way, the results may be very good. Designs of this sort have been lacking because the new students in art give no

attention to the work of the old-timers in their
own localities.

Fig. 108. Design for Remodeling and Veneering Old-Style Frame
Fronts.

The accompanying design, Fig. 108, was
made for a typical front in the old style of carpentry work. The ceilings of the main story

were very high. The basement was a full story. An outside flight of steps led to the front door on the second-floor level.

The new entrance was to be brought down to within a few feet of the grade. A half-flight of stairs was to lead from the vestibule down to the dining room, and another half-flight up to the parlor floor.

Heavy mouldings around the windows, of course, had to be removed. It was also necessary to see that foundations were good, and the walls and jambs plumb.

The design chosen for the remodeling calls for a veneer of glazed or enameled tile. Such a front would lead the fashion for a long time to come. The designer introduced a number of years ago enameled tile on exteriors, since which time it has become very popular. It is durable, always clean and cheerfully light, besides being elegant in appearance. This makes it well worth the expense.

The dimensions of the diagrams are suitable also for Roman brick, flatwise, so that that material can be used with terra-cotta mouldings. These mouldings are so constructed that a thin veneer may be used.

The veneering may also be of cast concrete blocks of the same sizes. It is this choice of materials and adaptability of these same forms to different dimensions, that makes the design economically practical.

The V-shaped vertical members of terra-

cotta, Fig. 109, stiffen the veneer at the corners
and windows. These are anchored to the wall,
and overlap the wall covering, holding it in
place. This also covers the ragged joint where
the tile has to be clipped. This is shown in the
section of the first and second story wall.

The basement is in Richardson courses;
only, instead of stone, the brick is laid alternately flat and edgewise. The vacant space is
filled with concrete, as shown in the section
(Fig. 109).

It is not enough to drive spikes into the wall
for anchors. They must hook firmly around the
sheathing or another spike at right angles. The
window head requires a small angle as shown
on the diagram. It must be bolted to the framework.

The space between the veneer and the siding
should be not less than an inch, and well filled
with cement mortar.

The frieze can be made of plain veneer, or
constructed of cement plaster on wire lath.

A plaster frieze can be modeled with good
effect. It should be done offhand while being
laid onto the metal lath. The centers and group-
ing of the pattern should first be indicated as
shown on the elevation. Then the carving can
be quickly modeled from a sketch. It is not
necessary to have all the details correspond
exactly if the symmetrical outline is preserved.
Fine detail should be avoided.

A tendency to pull down old buildings that

are substantial, in order to build up new ones,
depreciates the value of building improvements,

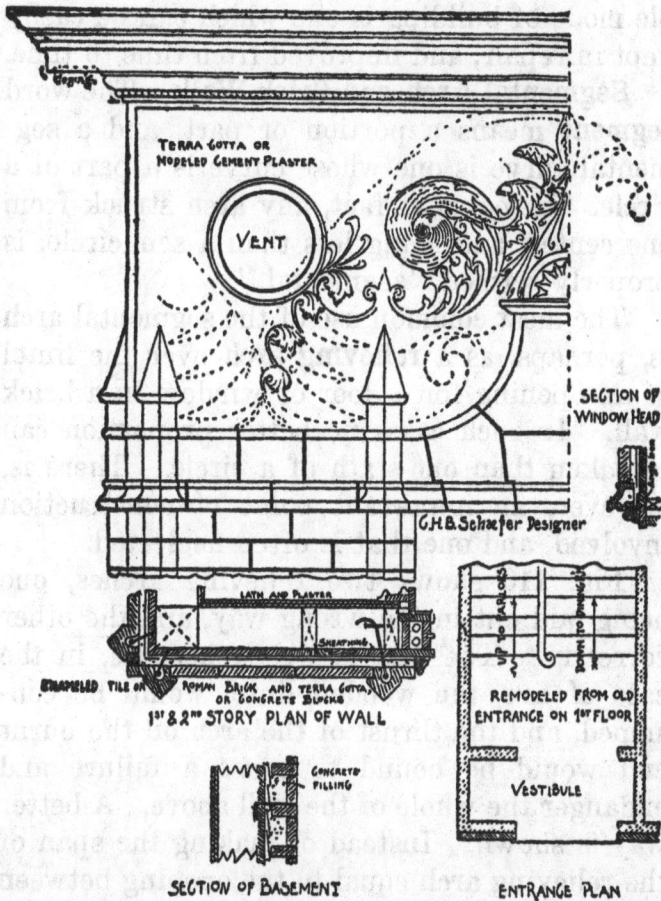

Fig. 109. Design and Construction for Tile Veneer Remodeling.

for it increases the expense in the long run,
enormously. Owners hesitate when such a prob-
ability presents itself in considering new work.

There is the owner's waste of time, incon-
venience, and, most of all, the neglect of other
affairs to be considered. Hence the most desirable mode of building is one which can be easily
kept in repair, and improved from time to time.

Segmental Arches in Brick Walls. The word
segment means a portion or part, and a segmental curve is one whose curve is a part of a
circle. In point of fact, any arch struck from
one center, and being less than a semicircle, is
properly termed "segmental."

The most common use of the segmental arch
is, perhaps, as a **relieving arch** over the lintel
of an opening for a door or window in a brick
wall. In such cases no better proportion can
be taken than one-sixth of a circle. There is,
however, an important point of construction
involved, and one that is often neglected.

Fig. 110 shows two relieving arches, one
being laid out in the wrong way, and the other
correctly. The first is wrong because, in the
case of fire, the wooden lintel would be consumed, and the thrust of the arch on the burnt
end would be bound to cause a failure and
endanger the whole of the wall above. A better
way is shown. Instead of making the span of
the relieving arch equal to the opening between
the jambs below, the arch springs from a point
over the extreme end of the wooden lintel. In
case of fire occurring and the lintel being
entirely consumed, the arch would be unaffected,
and would continue to carry the weight above.

Building inspectors and managers should insist on the adoption of this correct method, for it costs no more than the incorrect one, and the advantage of it in case of fire is greatly in its favor.

Of course, for such arches, no elaborate centering is necessary. The lintel is laid in position; and a piece of 1½-inch stuff is shaped to the curve of the arch, and laid upon the lintel to form the centering. The arch is then turned

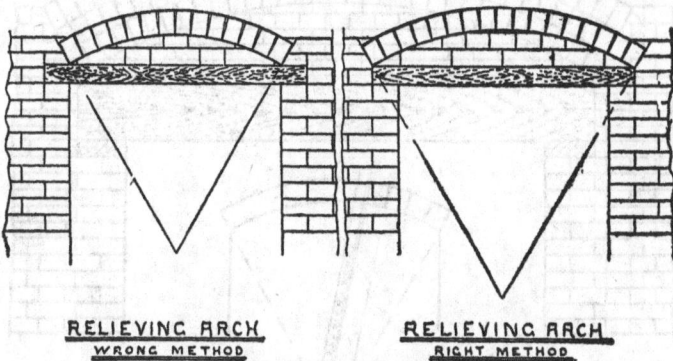

Fig. 110. Segmental Arches in Brick Walls.

upon this centering, which is removed when the mortar is properly set, the core being then filled in with brickwork.

For openings up to three feet or thereabouts, a relieving arch of a single ring of half-bricks is all that is required; but for larger openings, several rings may be used.

Fig. 111 shows an arch of three rings, and it will be noticed that each arch is separate and not bonded into its fellows. It will also

be noticed that the bricks of these rough relieving arches are not cut taper, and thus the joints
are slightly more open on the back of the arch
than on the under side. In making drawings
of such arches, the draftsman draws a ring
around the center from which the arch is struck,

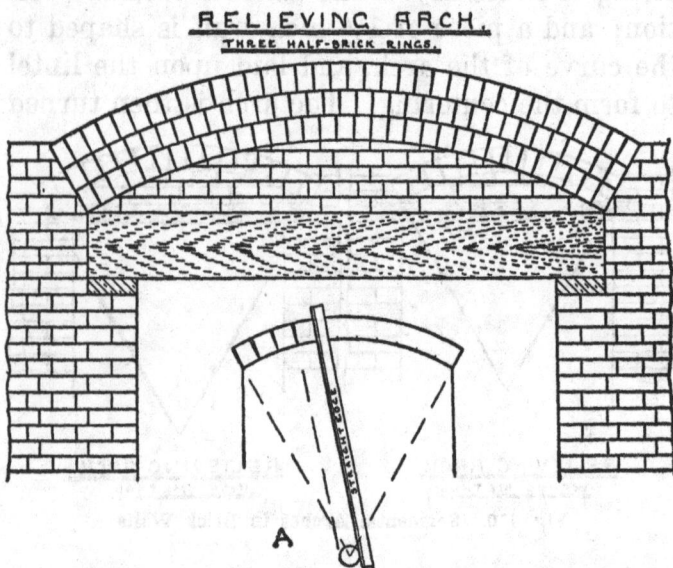

Fig. 111. Segmental Arch, and How to Lay It Out.

the diameter of the ring being the thickness of
the brick. This thickness is then stepped off
on the under side (soffit) of the arch with a
pair of dividers, and the straight edge placed
against the ring and one of the divisions on the
soffit (see **A**, Fig. 111).

How to Lay Out Arches. The chief problems, however, with which the practical layer-

out of arches is confronted, arise in connection
with the modern use of fine pressed brick for
so many first-class structures. For while the
mere curve is sufficient for practical purposes in
rough relieving arches, the arch made of facing
bricks, and forming a feature of some fine front,
must be set out exactly for the purpose of cutting and fitting, or perhaps moulding, the bricks
of which it is to be composed. Brick arches in
which the bricks have been specially cut or
moulded are generally termed **gauged arches**,
and are frequently used nowadays.

The **radius** of the arch is scarcely ever given
by the architect, the **rise** being almost invari-
ably denoted instead. The writer has before
him an elevation of a brick-fronted building
with some eight or ten openings of varying
widths, but the same rise is specified for all
the arches over them. This means that the
layer-out has to find the centers of the several
curves from the given particulars of their rise
and span. This he does as shown in Fig. 112,
the first being the geometrical method of the
drafting room; the second, the practical method
of the laying-out shop. In both cases, the center
from which the arch is struck is found at the
intersection of the lines drawn from the center
of each half of the arch.

As the bricks in gauged arches are used full
length, the thickness of the brick is marked off
around the back of the arch, and the joints
drawn to the center, as in Fig. 113, at left. The

joints are very fine, being usually specified to
be not more than ⅛ inch, the mortar being
either fine cement or lime putty.

In Europe, special bricks are made for such
arches, and are known as **red rubbers**. When
new, they are quite soft, and can be sawed with
a handsaw, and rubbed upon a block with sand
and water to form close joints. After being

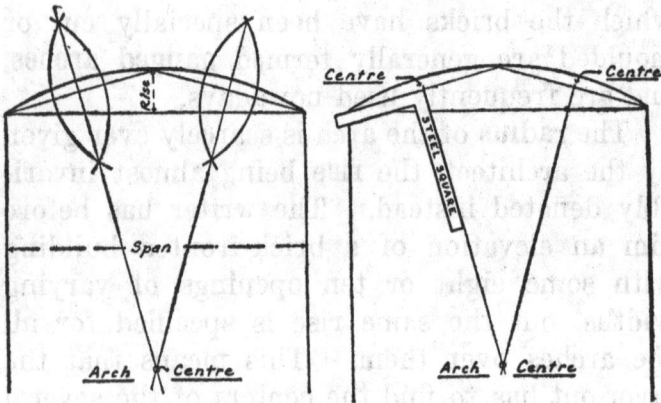

Fig. 112. To Find the Arch Centers.

exposed to the air for a time, the surface of these
bricks becomes exceedingly hard and impervious to the action of the weather. For the
red brick dwellings of "Queen Anne" and
"Colonial" style, now so much in vogue again,
such bricks are exceedingly useful. Not only
can they be cut for the characteristic flat arches
of these styles, but mouldings can be worked
on the angles, and panels formed to relieve
broad surfaces of wall. More often, however.

bricks for gauged arches are specially moulded to the builder's drawings by the makers of the face bricks, with fairly good results in the finished work.

The flat arch just referred to is also much used in brick fronts in city buildings, and is drawn as shown in Fig. 113, at right. It presents no difficulty to the layer-out, the joints

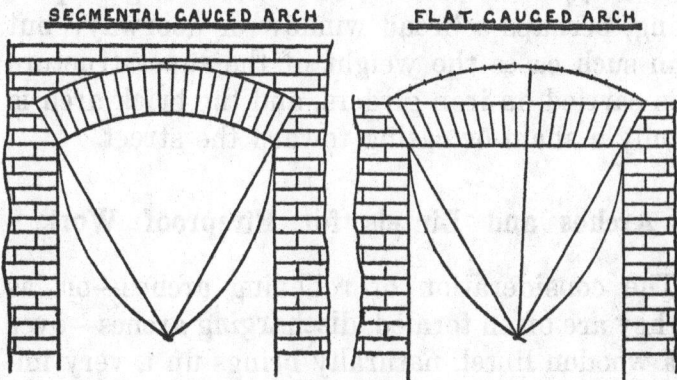

Fig. 113. Brick Arches.

being found by making a curve above the arch and stepping off the thickness of the bricks upon it.

There is one important point to be remembered, though, in building such arches—namely, that **a perfectly straight soffit will always appear to be sagging**. The remedy for this is to allow a trifling rise—say ½ inch for every three feet of span—which will be sufficient to make the under side of a flat arch look straight. This can be easily done on the job by laying

two strips tapering from nothing at the ends
to the required allowance at the middle, upon
the support or centering on which the mason
forms his arch.

Of course, flat arches are not very desirable,
from a structural standpoint, and should not be
used for spans more than four or five feet at
the outside. Occasionally, for the sake of uni-
formity, a flat arch is used over a larger opening, perhaps a broad window or doorway; but
in such cases the weight of the superstructure
is carried on iron girders, and the brick arch is
only a sham or casing toward the street.

Arches and Lintels for Fireproof Work.

The consideration of relieving arches—or, as
they are often termed, **discharging** arches—over
a wooden lintel, naturally brings up a very important question. In these days of fireproof
construction, when wood is being eliminated
from the structural parts of buildings wherever
possible, wooden lintels are not used in the best
practice. Instead of wood, iron I-beams and
artificial stone lintels are now largely employed
in the best class of work. But both of these
have one defect—namely, that, from their
nature, it is impossible to nail grounds or other
wooden finish to them. As the chief purpose of
a lintel is, of course, to form a square head for
a window or door-frame, this consideration is
important.

One of the best methods suggested for over-
coming this little difficulty—and not only suggested, but widely used in some parts of
Europe—is to form the lintel of coke breeze concrete. **Breeze** is the English term for the small
cinders left from the fires used for burning
bricks in a kiln, or from the manufacture of
coke in gas ovens. Mixed with cement, it makes
a concrete that is fireproof and that possesses
another useful quality in that nails may be
driven into it with ease. This last property has
led to the use of coke breeze in a variety of
places where wood was formerly employed.

For instance, thirty or forty years ago it
was still common to find bond timbers inserted
in brick and stone walls at such heights as would
render them convenient for fixing the trimmings
and finishing woodwork to afterwards. These
bond timbers were certain to shrink and were
also very liable to rot; and, as either of these
contingencies made them a source of possible
weakness to the wall, their use was gradually
discontinued, until it has ceased altogether.

The substitute at first proposed for the bond
timbers, and largely used for many years, was
a thin strip of wood built into a joint of the
masonry or brickwork, thus reducing the possibility of failure from shrinkage. The strips
were, however, in a great measure open to the
same objections as the larger bond timbers, and
many architects refused to allow them to be
used. Instead, they specified that hardwood

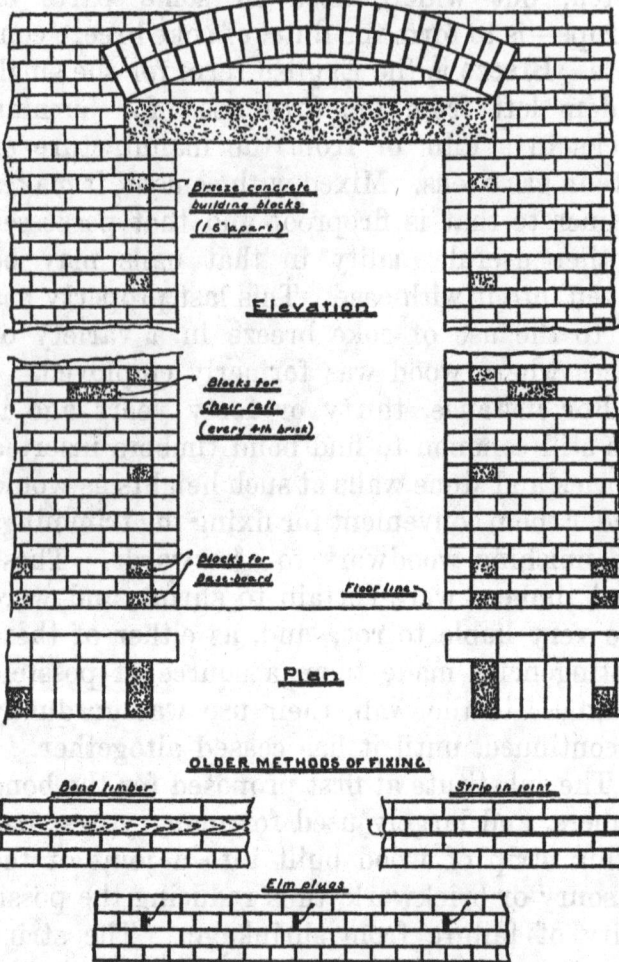

Fig. 114. Details of Fireproof Brick Construction.

(usually elm, on account of its non-liability to
split) plugs or wedges should be driven into the

joints at intervals, after the walls had set, the grounds or other woodwork being nailed to these plugs.

The invention of coke breeze concrete has, however, made it possible to substitute bricks and blocks made of it, which can be built into the wall during construction. They neither shrink nor rot, while their property of taking nails readily makes them ideal for the purpose of fixing woodwork.

Fig. 114 shows the application of coke breeze in the lintel and fixing blocks around a revealed opening for a doorway in a 12-inch wall. The blocks are also shown at each side, where they would be continued to form a fixing for the grounds for the chair rail and base-board. These blocks should be of the size of the bricks used in the building, and may, of course, be readily moulded in any suitable machine. Long before the use of machines for this purpose, however, many hundreds of such bricks were made in the roughest of wooden moulds improvised for the purpose. As coke slack or breeze is very often merely a nuisance to the gas manufacturer, its cost is next to nothing in many instances.

Fireproof Floors. The question of fireproof construction is by no means a new one, although the builders of the present generation have probably seen more attention devoted to it than did their predecessors. As showing an interesting method of dealing with the lintel problem, the upper portion of Fig. 115 is worthy of attention. A flat or camber arch takes the place of the wooden lintel, and is supported by means of an iron rod from the crown of the relieving arch above. For first-class work it would be hard to surpass this scheme, but its cost would prevent its adoption in anything but the very best practice.

While on the subject of fireproofing, another useful application of coke breeze concrete may be given. The many forms of iron and concrete fireproof floors are, in the majority of cases, covered with wood blocks or battens for the surface. Wood block floors are usually laid right on the concrete of the fireproof construction, some pitch or tar compound being used as a bedding cement. When batten floors are used, however, strips of quartering are usually laid on, or imbedded in the concrete, and the battens nailed to them. Many first-class architects, on the other hand, do not care for this method, and prefer that breeze concrete screeds should be laid in place of the quartering. If the screeds are run when the concrete below is still damp, they become an integral part of the body of the floor, and make a sounder job altogether.

The lower portion of Fig. 115 shows a simple rolled iron beam and concrete floor of the type generally known as "Dennett's," with board floor over. **A** shows a cross-section through the iron beams; and **B**, a cross-section through the concrete screeds, the boards being nailed to these as suggested.

FIREPROOF CONSTRUCTION
A GAUGED RELIEVING ARCH OVER A FLAT OR CAMBER ARCH
WITH SUPPORTING ROD

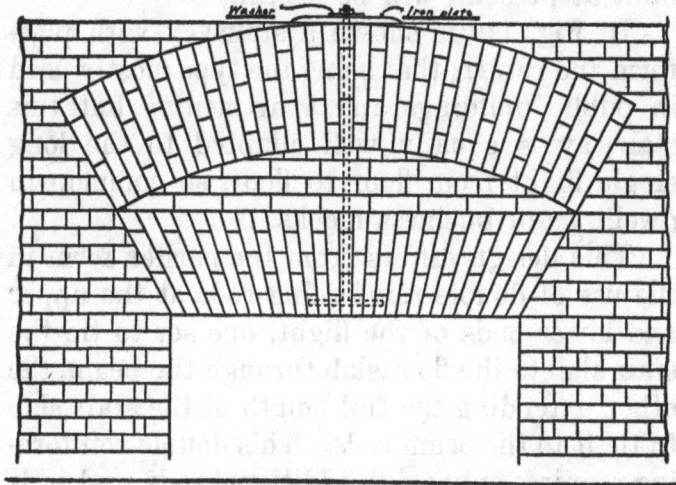

SIMPLE FIREPROOF FLOOR
SHOWING USE OF COKE BREEZE CONCRETE SCREEDS

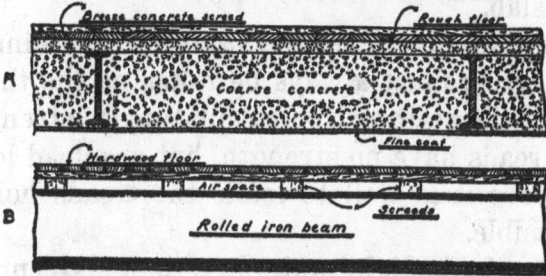

Fig. 115. Details of Fireproof Construction.

Reinforced Concrete Stairs. Concrete for stairs is rapidly taking the place of other materials in buildings of every kind. A few simple instructions as to the stair load and some remarks concerning the weak points of stairs built several years ago, together with a description of the reinforced concrete stairway as it is built at present, will be useful.

In Fig. 116 is shown a stairway with platform and return, that being the type mostly used in public buildings and apartments; but this stairway is equally well adapted to the long single flight from floor to floor, so common in dwellings or business blocks.

This design differs from the earlier form in the use of double reinforcing rods at the upper and lower ends of the flight, one set to tie the stair slab to the floor slab through the beam, the other extending the full length of the stair slab to tie it to the beam rods. This double reinforcing requires only a few additional rods, and adds much strength to the weakest spot of the stair slab.

The thickness of the stair at its thinnest place, **A A**, must always be taken as the thickness of the stair slab in calculating its strength. The treads have no strength, but are dead load; therefore it is well to make the treads hollow if possible.

For single-flight stairs less than sixteen feet long and less than five feet wide, it is well to use a four-inch slab, unless for very heavy loads, and support it so that no two supports are more than ten feet apart.

This slab is reinforced with half-inch square twisted or five-eighths round rods, spaced as follows—for dwellings, ten inch centers; for busi-

ness blocks, six to eight inch centers; and for
public halls and factories, two to four inch centers. These rods have one inch of concrete
under them. At every tread is a cross-rod of
same weight tied to each with No. 10 wire.

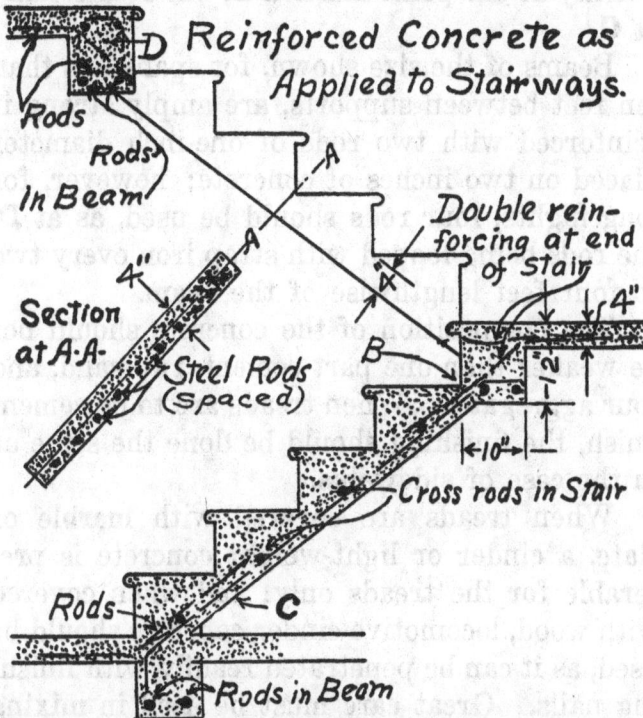

Fig. 116. Details, Concrete Fireproof Stairs.

The full-length rods all pass over or hook
onto the rods in the cross-beams; and the double
reinforcing consists of rods about five feet long,
passing over the beam, with the ends projecting
equally into the stair and floor slabs.

An examination of concrete stairs built without this additional reinforcing, has in several instances revealed the fact that cracks, if any, appear near the ends of the slab; in fact, within ten inches of the supporting beams, most frequently at the point marked **B**, but occasionally at **C**.

Beams of the size shown, for spans less than ten feet between supports, are amply strong if reinforced with two rods of one-inch diameter placed on two inches of concrete; however, for long flights, four rods should be used, as at **D**, the rods being looped with strap iron every two to four feet lengthwise of the beam.

The composition of the concrete should not be weaker than one part cement, two sand, and four aggregates. When treads are to be cement finish, the finishing should be done the same as in the case of sidewalks.

When treads are covered with marble or slate, a cinder or light-weight concrete is preferable for the treads only; and when covered with wood, locomotive cinder concrete should be used, as it can be penetrated readily with finishing nails. Great care must be used in mixing and tamping concrete for stair work.

How to Apply the Wood Trim. There are a number of features in connection with modern fireproof construction that are of particular interest to carpenters, especially to those working in the larger cities where steel, tile, and cement are now so generally used. A question

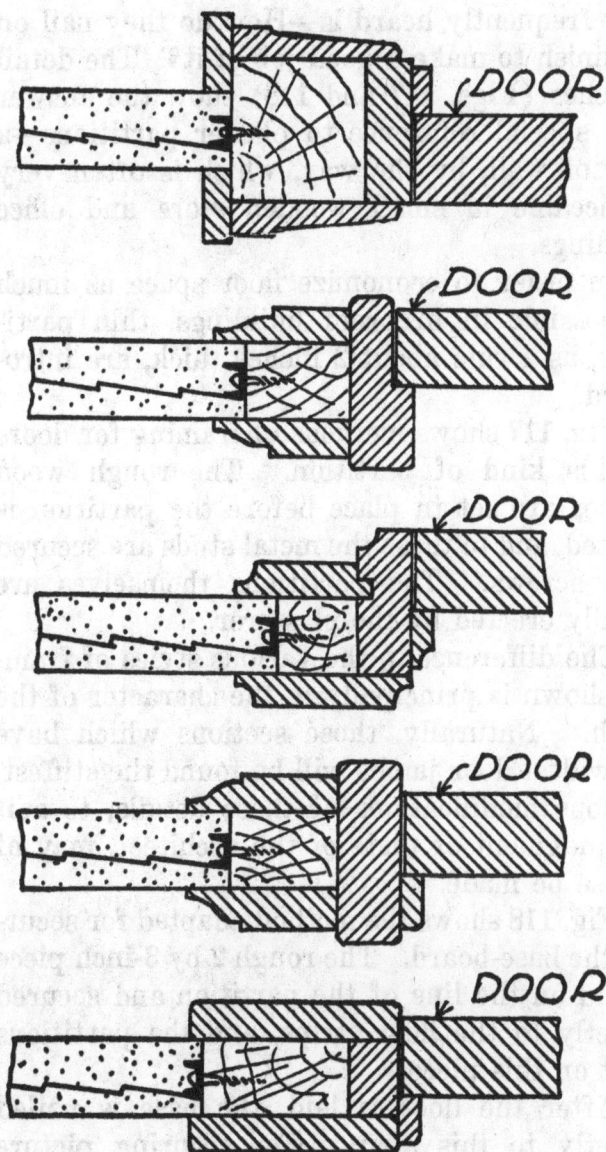

Fig. 117. Door Framing in Plaster Partitions.

very frequently heard is—How do they nail on the finish to make a good job of it? The detail sketches (Figs. 117 and 118) show the answer with special reference to plaster partitions—a type of wall, by the way, which is often very serviceable in small wooden store and office buildings.

In order to economize floor space as much as possible in fireproof buildings, thin partitions, as a rule about 2 inches thick, are introduced.

Fig. 117 shows methods of framing for doors in this kind of partition. The rough wood frames are set in place before the partition is erected, and to these the metal studs are secured with screws. The partitions themselves are usually erected by the plasterer.

The difference in the various styles of framing shown is principally in the character of the finish. Naturally, those sections which have the widest door jambs will be found the stiffest. Various modifications of these details, to suit the judgment or taste of the architect, may of course be made.

Fig. 118 shows the method adapted for securing the base-board. The rough 2 by 3-inch piece is laid on the line of the partition and secured directly to the floor strips, and the partitions built on this piece.

After the floor is laid, the base is nailed directly to this strip. For securing picture moulding, strips of wood may be laced to the metal lath at the required height, before the plastering is done. These are sufficiently firm, after the plaster has dried, to hold the picture

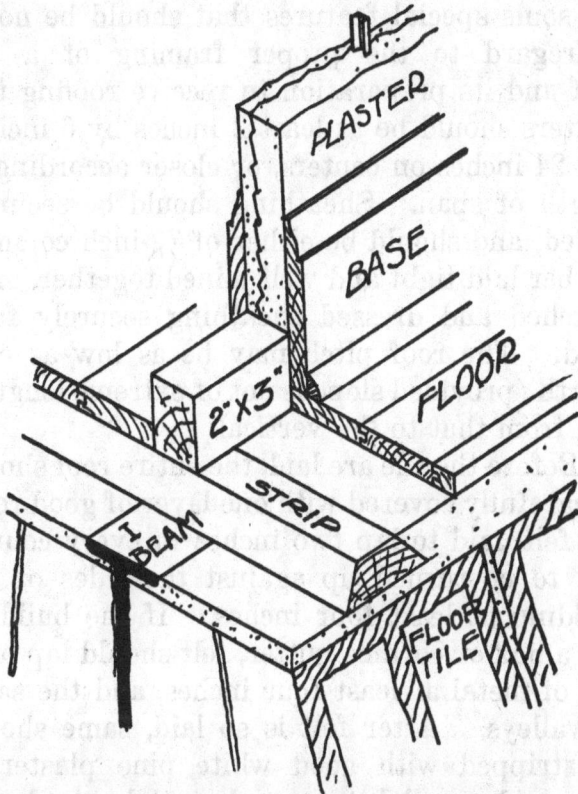

Fig. 118. Baseboard and Floor Construction.

moulding, which should be put up with screws instead of nails. In case close-warp metal lathing is used, the screws will engage the meshes of wire work sufficiently to hold the picture moulding without any preliminary strips.

With slight modifications, the methods shown for door framing could be adapted to hollow tile partitions.

Framing for Heavy Roofing—Tile. There are some special features that should be noted in regard to the proper framing of a tile roof and its preparation to receive roofing tile. Rafters should be at least 2 inches by 6 inches, and 24 inches on centers, or closer according to length of span. Sheathing should be securely nailed, and should be either of $\frac{7}{8}$-inch common lumber laid tight and well joined together, or of matched and dressed sheathing securely fastened. The roof pitch may be as low as one-fourth (provided slope is not of extreme length), and from that to the vertical.

Before the tile are laid, the entire roof should be carefully covered with one layer of good roofing felt, laid to lap two inches in every course, and to be turned up against the sides of the building at least four inches. If the building has a box or cornice gutter, felt should lap over top of metal at least four inches, and the same at valleys. After felt is so laid, same should be stripped with good white pine plastering lath, laid parallel, true, and straight, to facing board at eaves. The top edge of first line of lath should be 12 inches above the lower edge of facing board or starting strip; and thereafter not less than 12 inches nor more than $12\frac{1}{4}$ inches space allowed from the top edge of each line of lath to the top edge of the next above and parallel. The tile hook over these strips; and each tile is fastened with a seven-penny galvanized or copper wire nail.

All ridge-boards should extend three inches above top of sheathing, and hip boards two and one-half inches, and both be of seven-eighths-inch common lumber. Facing board or starting strips at eaves under bottom end of tile will extend up above the top edge of sheathing one and five-eighths inches. In all cases facing boards at gable ends should be flush with the sheathing.

In some cases an **open** roof construction is used—that is, one with no sheathing under the tile. In that case, there must be a space of twelve inches between the lower edge of the lowest purlin to top edge of the purlin next above it, and thereafter a space of not less than twelve inches nor more than twelve and one-fourth inches between the top edge of each purlin to the top edge of the purlin next above it. These purlin strips should be $7/8$ inch by 2 inches or over, the bottom strip $1\frac{1}{4}$ inches higher than the strip next above it—that is, $2\frac{3}{8}$ inches by $7/8$ inch. In this construction the hip and ridge strips should be the same as if the building were sheathed.

Framing for Slate Roofs. It is the prevailing opinion of people not familiar with the use of slate for roofing purposes, that a building should be constructed very much stronger for slate than for other roofing materials. This is a mistake, as any building strong enough for shingles, tin, or iron is strong enough for slate. Two-by-six rafters, eighteen feet long, two feet from centers, give all the strength necessary. The writer has seen hundreds of houses roofed with slate where the rafters were two-by-four, two feet from centers, sixteen feet long, with collar beam nailed across one-third of the way down from the top.

Slate can be depended upon to make a roof perfectly water-tight on any pitch down to one-fifth. Half-pitch or steeper makes the best roof both for looks and strength, as it throws the weight on the walls more than on the rafters, and causes the snow to slide off clean, thereby never overloading any one part of the roof.

Matched lumber is best for sheathing for any roof; but surfaced boards from six inches to ten inches wide make a good job, and are used on a large majority of the buildings now being put up. Sheathing boards, when not matched, should be nailed at both edges on rafters, which should not be over two feet apart. Wide boards, when used for sheathing, are liable to warp and curl up at the edge, thus affecting the slate. While it may not break the slate, it raises the courses, marring the appearance of the roof. Very often a roof that lies well and smooth when done, apparently gets rough and the slates stick up. The roofer is often blamed for this when the cause is really in the sheathing. Great care should be used in putting on the sheathing, that there are no lumps or uneven thicknesses in the boards, as they will surely show after the slate is put on. This especially applies on curved roofs or round towers, dormers, etc. In all such, the rafters should be close together and the sheathing perfectly solid and smooth. Where the sheathing is not solid it is almost impossible to make a good smooth job, for the reason that in driving one nail it jars the next slate loose.

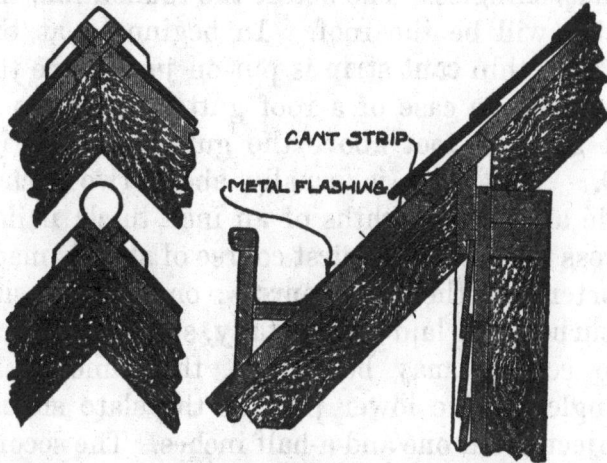

Fig. 119. Proper Construction at Ridge and Eaves for Slate.

Lath or strips are often used instead of sheathing boards on which to lay slate; but in such case the lath should be at least one and one-quarter by two and one-half inches, and must be spaced to suit the size of slate used. They should be placed so that the upper end of each slate will rest in the center of the lath. This plan is a good one for barn roofs, as it allows some ventilation between the slate; but where a perfectly snow-tight roof is wanted, the slate should be pointed with hair mortar on the under side of the slate at the upper end of each course, also at the joints between the slate.

Tarred or other waterproofed paper should be used under slate where the same is laid on sheathing boards. This will insure a roof perfectly tight against fine snow.

Slate roofs require about the same foundation as shingles. The better the foundation, the better will be the roof. In beginning at the eaves, a thin **cant strip** is put on just above the eaves; or, in case of a roof gutter, the strip is put about a foot above the gutter, as in Fig. 119. This strip is usually about two inches wide and three-eighths of an inch thick, nailed across the roof. The first course of slate is made shorter than the other courses; or the usual size is turned and laid horizontally, so that the first two courses may be double, the same as in shingles. The lower part of the slate should project about one and a-half inches. The second course up should lap about three inches over the first or double course. When nearing the peak, the lap may be varied a little to make the slate come out right.

Barn Framing

Next to house framing of the various kinds already discussed, the branch of construction most important for the practical carpenter, as offering itself most frequently in actual work, is **barn framing**. There are building contractors who make a specialty of this kind of work; and their advice on numerous special points—which we have been able to gather and here present—may be taken as authoritative. Barn framing, for itself and because of its relation to other kinds of timber framing, should be well understood.

At the present time there are two general types of timber-framed barns in use. The first are **heavy timber barns,** all structural members being heavy, squared timbers, with joints tenoned and pinned—a survival of the sturdy construction of earlier days when timber was plentiful and was all hewed out by hand; and the second, **plank-framed barns,** claimed by many to mark a decided improvement over the other as being more easily erected, just as substantial, and decidedly cheaper, especially where large timber is not easily secured.

The principles and important features of these types of barn construction, together with some considerations common to them both, will now be given.

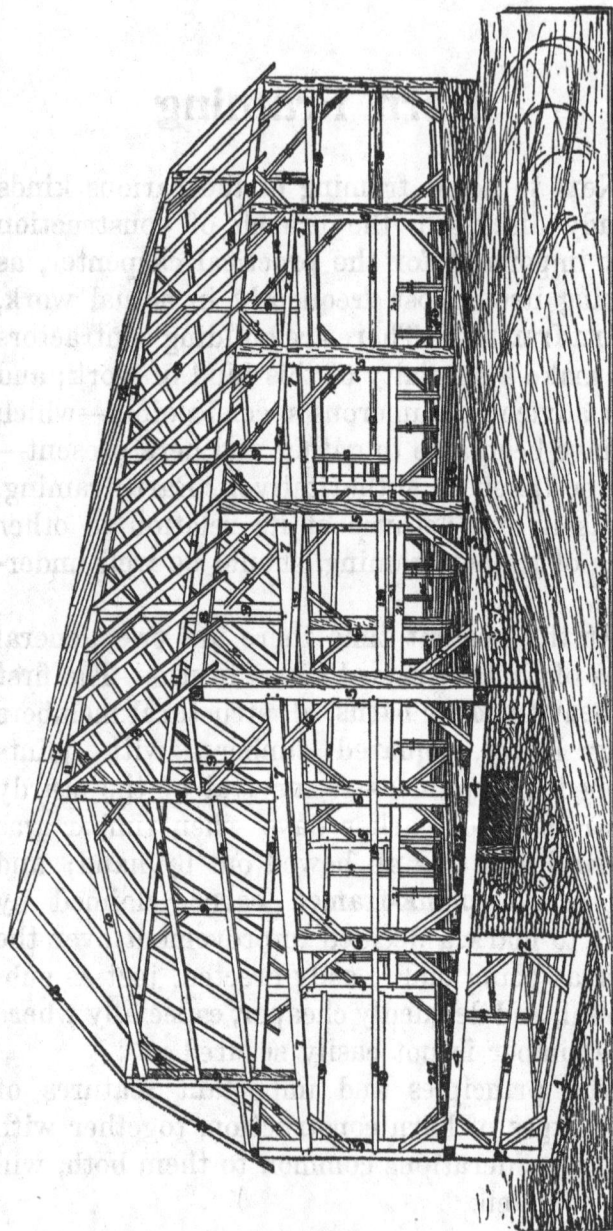

Fig. 120. General Framing Plan for a Heavy Timber Barn.

HEAVY TIMBER BARNS

General Framing Plan. Fig. 120 shows the
general plan of a barn frame of a building 38
by 64 feet, with 16-foot main posts, gambrel
roof, basement, and double driveway floors. It
is typical heavy timber construction, with boxed
mortise-and-tenon joints.

One point worthy of note for storage barns
such as this, is the effort to get height—not
width—to get as much under a given amount
of roof as possible. That this is an advantage,
has been slowly but well learned; the farmer
will tell you that the bay or mow is only half-
full when filled even with the main plates. The
weight of the portion to be stored above keeps
pressing the lower part down—an advantage
that is lost in the low, wide barn. All the crops
are lifted by means of rope and tackle, and
carried by means of a steel track and car the
full length of the barn just under the ridge of
the roof, and then dropped to their proper places
in the bays, mows, or the loft, over the driveway floors, the power being furnished by a horse
team.

Another important principle in the construction of these buildings is that no timbers supporting a heavy load—as beams, cross-sills,
etc.—are made to rest on their tenons alone,
but all have a shoulder or bearing across the
whole end of the stick.

The drawing shows the ends of beams boxed

into the posts, while the ends of the cross-sills have a bearing on the end of the basement posts below. Good framing requires also that no braces shall hold against the tenon alone, but they should have a boxing of at least one-half inch in depth.

Another important principle is that of **draw-bore pinning**. By this is meant pins that hold the tenons; ordinary drift bolts or spikes would not do, as a pin made of wood is larger than the iron ones would be, and will not side-cut or press into the wood so much as the iron ones would do under a heavy load. Also, the pins must have a long taper, to give **draw**. The draw-bore holes in the tenon and those through the mortise are given one-eighth inch draw or variation in such direction as will tend to pull the tenon and seizing tighter into the mortise and boxing. That is to say, the distance from the joint to the draw-bore hole is one-eighth inch greater in the mortise than in the tenon.

All of the large cross-timbers should be of whole sticks—not spliced—that is, the beams and cross-sills in particular; but the main sills and plates may be spliced at every bent or post if desired. Sometimes long timbers are very hard to get, in which case the width of the barn can be made to conform to length of timbers obtainable, from 30 to 40 feet.

Some people have a notion that several planks spiked together will make a stronger stick of timber than a solid one. This is not

true; after having tried it many times, the
writer is prepared to say that you can make a
stick fairly stiff one way; but, with all the
spikes you can drive, it cannot be made as stiff
the other way.

The length and number of posts in such a
barn are entirely arbitrary, depending on the

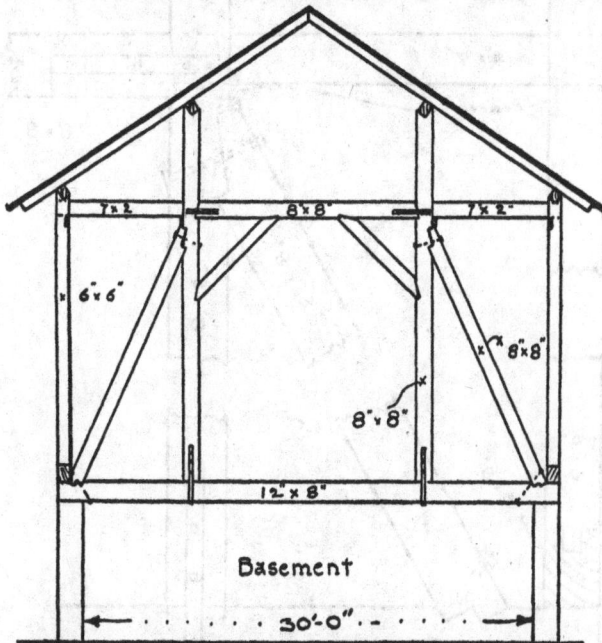

Fig. 121. Truss for Barn without Basement Posts.

size of barn desired. The height of the purlin
plates is also variable.

This kind of barn is built to best advantage
on a gentle hillside, thus allowing an easy driveway to the main floor of the barn, and giving
opportunity for a basement underneath to

afford valuable storage room for implements;
this basement space also provides warm stables.

While there are other forms of heavy timber
construction which afford greater strength than

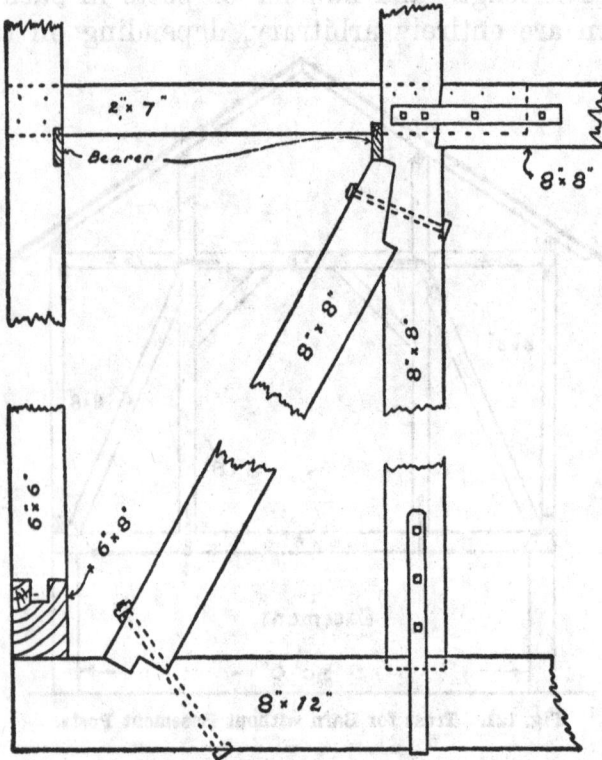

Fig. 122. Details of Heavy Framing.

that shown in the drawing, this method has been
found to answer every purpose; and the added
advantages of other methods of framing seldom
warrant the difference in cost.

At first glance at the construction shown in Fig. 120, one would think such a frame a wilderness of timbers. As a matter of fact, however, the system is simple, and the number of names of different members or parts is not great. The accompanying list gives the names and sizes of the different numbered parts:

PARTS OF A HEAVY TIMBER BARN.

	Name of Part	Size
1.	Basement sill	10 by 12 inches
2.	Basement posts	12 by 12 "
3.	Main sill	10 by 10 "
4.	Cross-sill	10 by 10 "
5.	Main post	8 by 8 "
6.	Center post	8 by 8 "
7.	Main beams	8 by 10 "
8.	Main plate	8 by 8 "
9.	Purlin posts	6 by 6 "
10.	Purlin beams	6 by 6 "
11.	Purlin plate	6 by 6 "
12.	Upper rafters	2 by 6 "
13.	Lower rafters	2 by 6 "
14.	Purlin girts	4 by 6 "
15.	Purlin braces	3 by 4 "
16.	3-ft.-run brace	3 by 4 "
16$\frac{1}{2}$.	2$\frac{1}{2}$-ft.-run brace	3 by 4 "
17.	3$\frac{1}{2}$-ft.-run brace	3 by 4 "
18.	End girts	4 by 6 "
19.	Side girts	4 by 6 "
20.	Door girts	4 by 6 "
21.	Breast girt	6 by 8 "
22.	Breast girt studs	3 by 4 "
23.	Ladder post	3 by 4 "
24.	Door posts	4 by 4 "
25.	Overlays, top and ends flatted to	6 "
26.	Sleepers	6 by 6 "

Details for Heavy Timber Framing. Fig. 121 is a sketch of a barn bent on an eight-foot basement, framed as a simple truss so as to do away with the posts in the basement. The roof and floor loads are supported, the raking members carrying the stress down to the side walls of basement.

Fig. 122 shows details of framing; which should be carefully done, as the truss has to bear moving loads as well as the weight of hay, animals, etc., and the roof itself. The sizes of

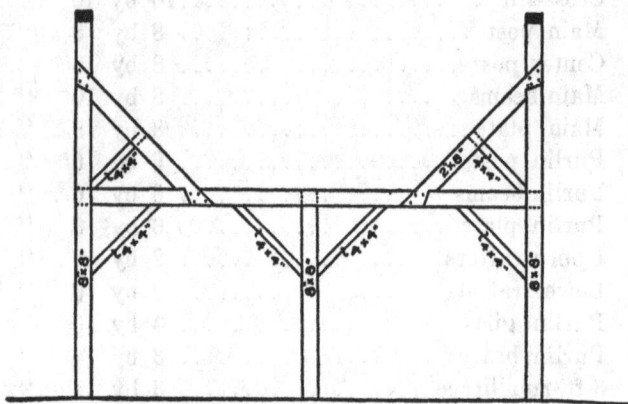

Fig. 123. Bracing to Prevent Outward Spreading.

timber given allow for trusses being about 8 feet apart.

Special bracing is sometimes necessary to prevent a large barn from spreading in the hay-mow. Some arrangements are all right for outside pressure, but not to withstand the hay pressure from the inside. Frequently barns give out in that way and have been fixed over; but it is not a very desirable job. Fig. 123 is a small sketch which shows how to make a stronger frame than the others, and with less braces. The long braces go in between the 2 by 8-inch braces, and should be well spiked. There should be braces between bents to keep the posts from twisting, in case the foundation should settle.

Strong Purlin Post Bracing. Fig. 124 shows a method of framing the center bents of a barn,

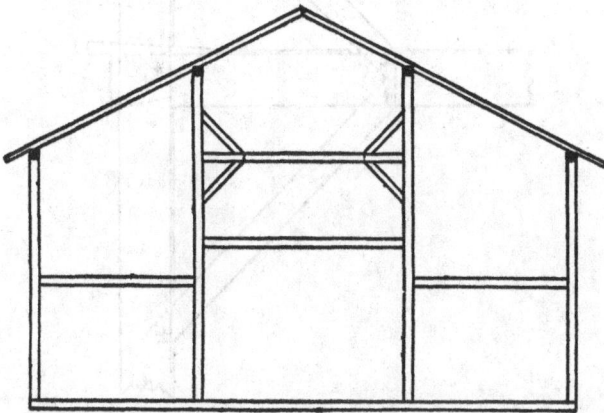

Fig. 124. Center Bent, Showing Purlin Post Bracing.

which is especially adapted to barns having flat roofs and equipped with hay carriers. A very strong bond is needed in such barns between the purlin posts, to prevent the excessive weight placed on the top rafters by the hay carrier from spreading them apart. The girder is also low

enough for the loaded hay carrier to pass
over it.

There is no girder between the top of the
shed post and the purlin; for, when the purlins
are framed as described, there is no need of

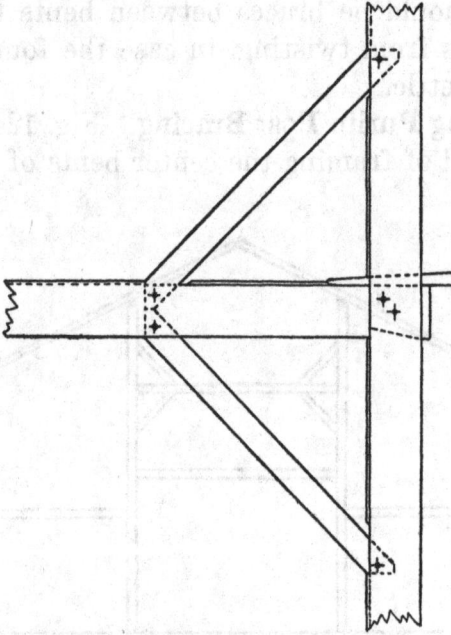

Fig. 125. Mortise Joint for Cross-Tie.

such a girder, as the rafters will hold the shed
post to its place if it has a tie at the bottom and
a loft tie up some 8 or 9 feet. Of course, where
the barn is extremely high, it is well to put in
this girder.

Fig. 125 shows a section of the purlin post,
center girder, and braces, drawn to a large scale

to illustrate the method of dovetail mortise joint
used.

Expert advice was asked not long ago concerning the best method of construction for a
barn somewhat similar to this one. It was to
be a barn for 50 acres of land; its dimensions
26 by 46 feet, and 18 feet to plate, with a solid
concrete wall under the entire barn. Would it
be advisable to use big timbers mortised and

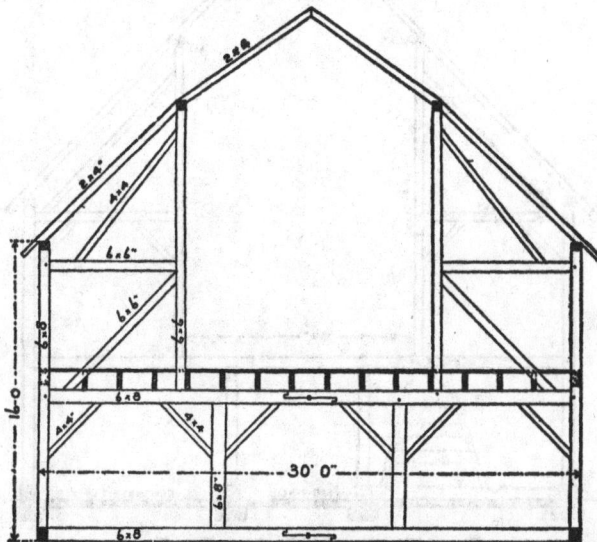

CROSS SECTION.
Fig. 126. Center Bent of Hay Barn.

tenoned together, or a balloon frame with 2 by
4's placed on 2-foot centers and sided with patented siding? How should it be tied together?
Would 2 by 8 inches, 26 feet long, be all right
with tight mow floor over all except driveway?

Will 2 by 6-inch rafters be all right for that
width to be used with a shingle roof?

These were the points inquired about. As
this barn is not large, it would not be necessary
to use heavy timbers, except for the sills; and
for this, not more than 6 by 8 inch, properly

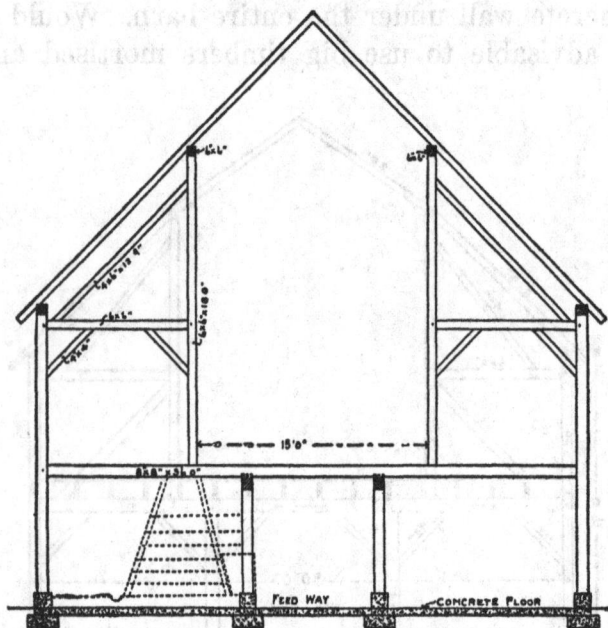

Fig. 127. Center Bent of Hay Barn.

framed and anchored to the concrete with bolts.
Two-by-four studding set on 24-inch centers
would be all right, but heavier than 2 by 8-inch
joist should be used for the hay-mow floor,
unless there is to be a support through or near
the center. About every third studding above

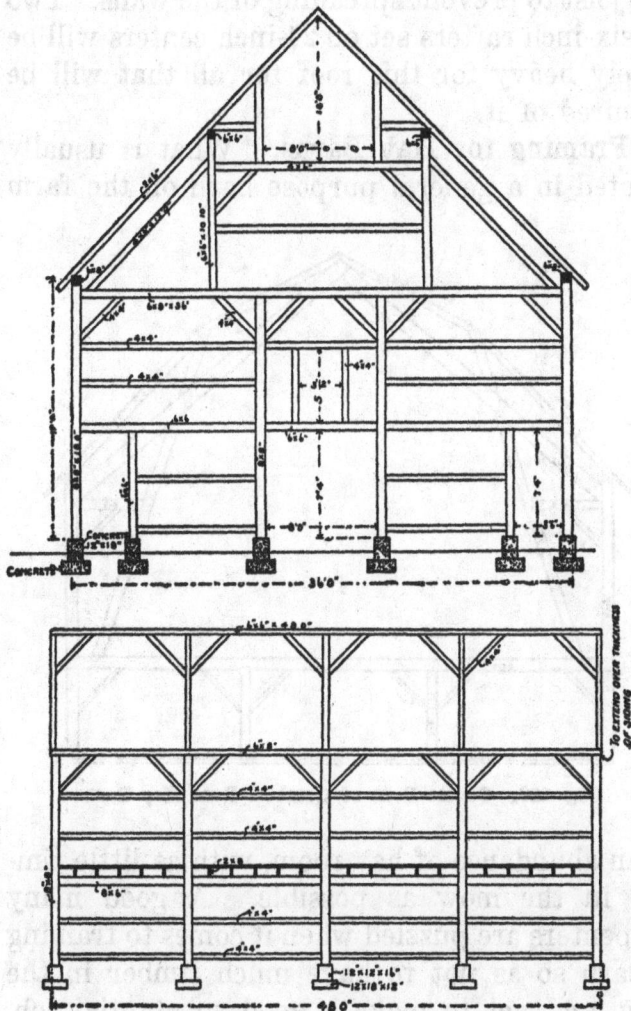

Fig. 128. End and Side Wall Framing for Hay Barn.

the hay-mow floor should be tied with braces to the joist to prevent spreading of the walls. Two by six-inch rafters set on 24-inch centers will be amply heavy for this roof for all that will be required of it.

Framing for Hay Barns. What is usually wanted in a general purpose barn on the farm

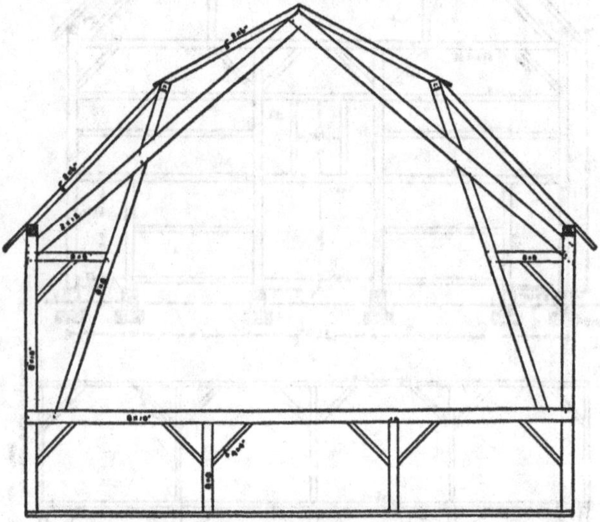

Fig. 129. Center Bent of Gambrel Roof Hay Barn.

is an abundance of hay room, with as little timber in the mow as possible. A good many carpenters are puzzled when it comes to framing a barn so as not to have much timber in the way, and yet to make a good substantial job. Fig. 126 shows good construction for a country barn that pleases all who see it. It is the center bent to which attention is called.

Figs. 127 and 128 show three sections of another barn well framed for hay storage. It is a good plan, having no timbers in the way if a fork is used in mowing hay. This is a barn very often built.

All floors are of concrete, and the posts rest on concrete blocks. All main posts are 8 by 8 inches; girders, 6 by 8 inches; plates, 6 by 8 inches; purlin plates, 6 by 6 inches; purlin posts,

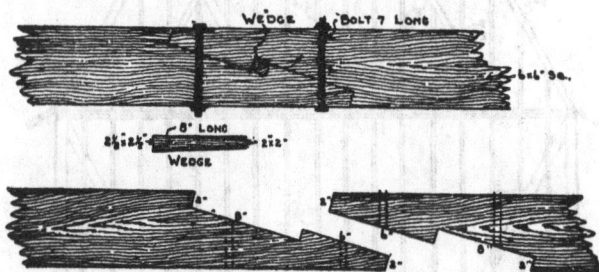

Fig. 130. Good Splice for Heavy Timbers.

6 by 6 inches; joist bearers, 6 by 8 inches; cross-ties, 6 by 6 inches; rail ties, braces, etc., 4 by 4 inches; rafters, 2 by 6 inches; joists, 2 by 8 inches, spaced two feet apart. The entire frame should be put together with mortise-and-tenon joints. This kind of barn should be sided with No. 1 barn siding, and cracks battened with bevel batten. The general dimensions are 36 feet by 48 feet. Fig. 129 shows the center bent of a Gambrel roof hay barn, 40 by 80 feet.

Splice for Heavy Timbers. It often happens that timbers of proper dimensions for a given purpose are not long enough and have to be spliced. Fig. 130 is a diagram of as good a spliced joint as there is. It is strong, durable, and easily made.

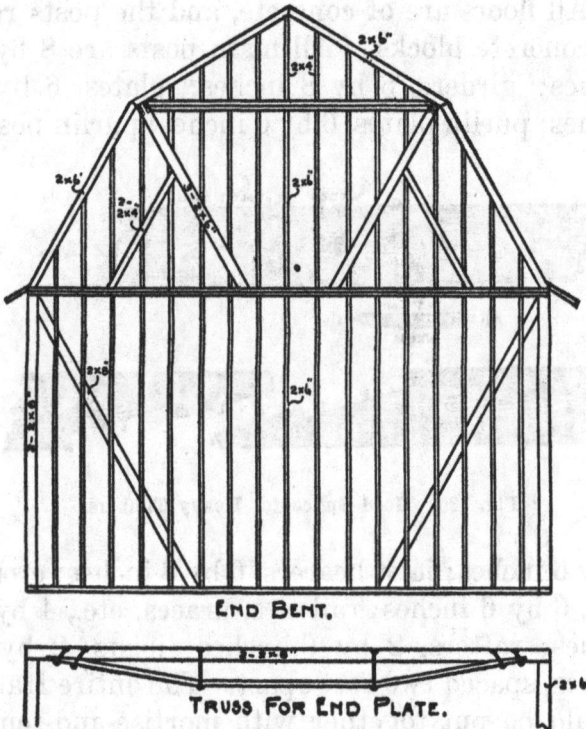

END BENT.

TRUSS FOR END PLATE.

Fig. 131. Plank-Frame Construction.

PLANK-FRAMED BARNS

With the scarcity of heavy timbers and consequent cost, it is time carpenters who are to erect barns should give some study to the newer methods of framing, where timber is from 6 to

12 inches wide, and none thicker than 2 inches.
The use of modern hay and grain elevating
machinery, calls for barns with open centers;
hence upper cross-ties, collar beams, etc., are
in the way, and they are quite unnecessary.

A plank frame of size and construction indi-

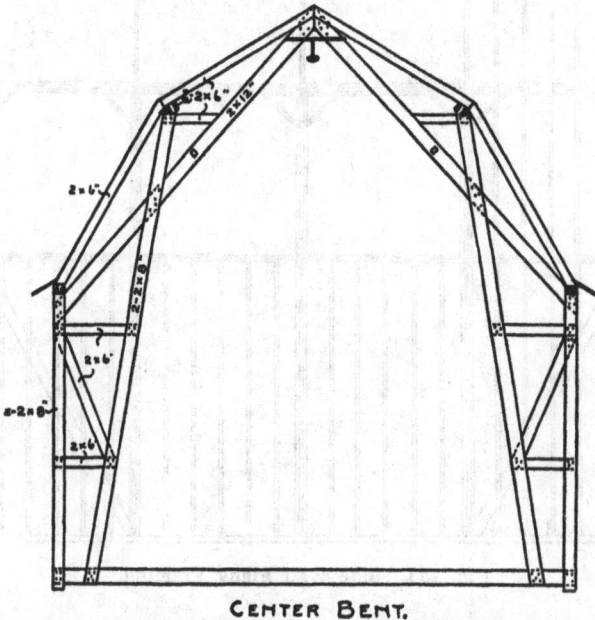

CENTER BENT.

Fig. 132. Wind Bracing for Plank-Frame Barn.

cated in Figs. 131, 132, and 133, is in every way
satisfactory, and is fully as strong as an old-
fashioned frame made of timbers 8 to 12 inches
square. It is about two-thirds as costly; and
less experienced carpenters are required to
erect it.

In this plank frame, there are no timbers larger than 2 by 12 inches, which are doubled and trebled where great strength is required. Where tensile strength is required, two 2 by 8-inch are nearly as good as an 8 by 8-inch tenoned and fastened in the old-fashioned way with a

Fig. 133. Side Wall Plank Framing.

pin.   In this frame there are no tenons, all members being put together with spikes.

Fig. 134 shows the method of **raising** or erecting the bents after they are framed.

The posts for this barn are 20 feet high, and are built up of joist spiked and bolted together. The bracing and cross-pieces were framed together in sections before raising to their position. The inner set of posts, set raking, extend up to form the purlin. The rafters, of which there are two sets, are placed on two-foot centers, lapping at this point.

This kind of construction is very popular in some sections of the country, and, when well built, makes a very substantial job. The timbers, being entirely of joist, are such as can be readily obtained from any lumber yard that is worthy of the name. This is a thing in itself that should not be overlooked, as solid framing timbers are not so plentiful as in years gone by, and in many places are not carried in stock.

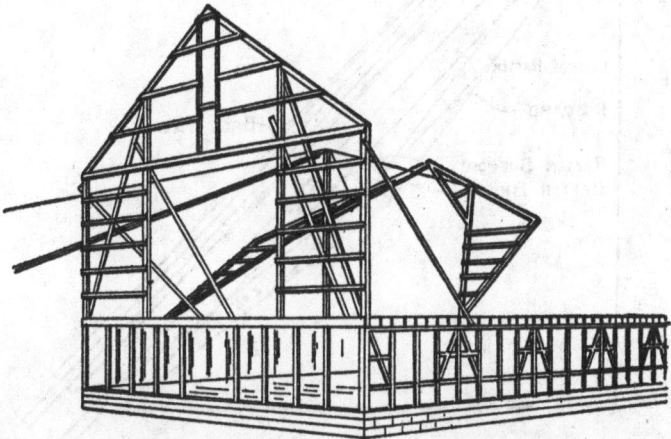

Fig. 134. How the Bents are Raised.

The dealer places the order after he receives the contract to furnish the lumber; and this means that the contractor must wait for the lumber to be shipped from some distant mill, and more

than likely to be first cut from the round timbers; and when it does finally arrive on the
building site, it is yet green and really not fit
to frame for the best results.

Another thing in favor of the plank frame
is that the timbers are light and more easily
put together than by the mortise-and-tenon
method.

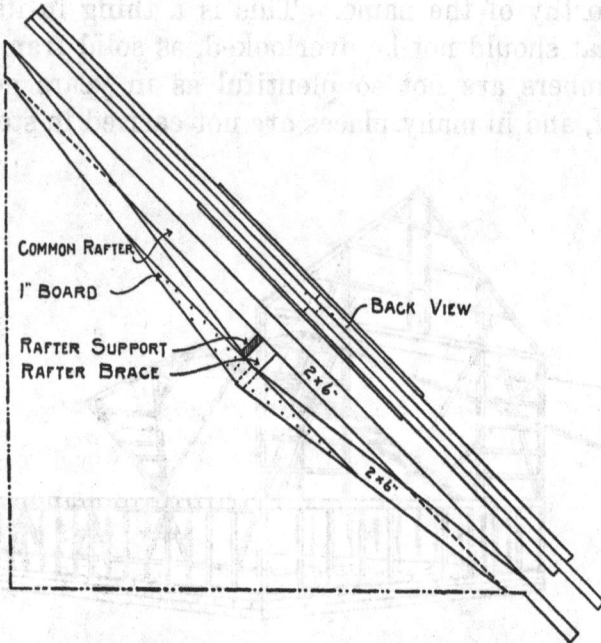

Fig. 135.   Queen Rafter for Roof Support.

Roof Supports. Fig. 135 is a sketch of what
is called a **queen rafter.** It is used for barns
up to 34 feet wide, without putting in purlins.

While it possesses considerable merit for stiffening the roof, there is nothing in it to keep the sides of the building from spreading, as it does not form a tie, which is necessary when the loft floor is far below the plate.

To secure the additional strength needed, the rafters should be set on 24-inch centers, and

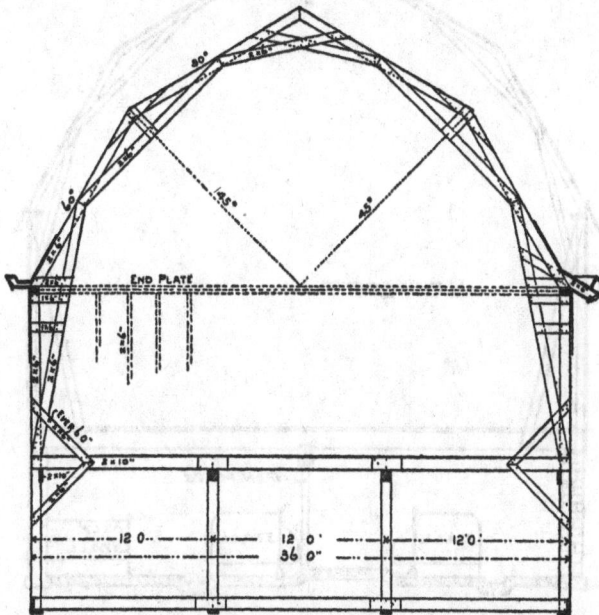

Fig. 136. Plank-Framed Truss for Gambrel Roof.

sheathing put on diagonally toward the center, close and well nailed. The center rafters where the sheathing meet should be doubled and well spiked together. The floor-joist lap onto each stud, and should be well nailed to prevent spreading.

Support for Gambrel Roof. To support a
gambrel roof on a barn 36 feet wide, having
posts 20 feet high, the floor being 11 feet below
the plate, requires extra strong bracing. A construction using 6-inch studding set on 24-inch

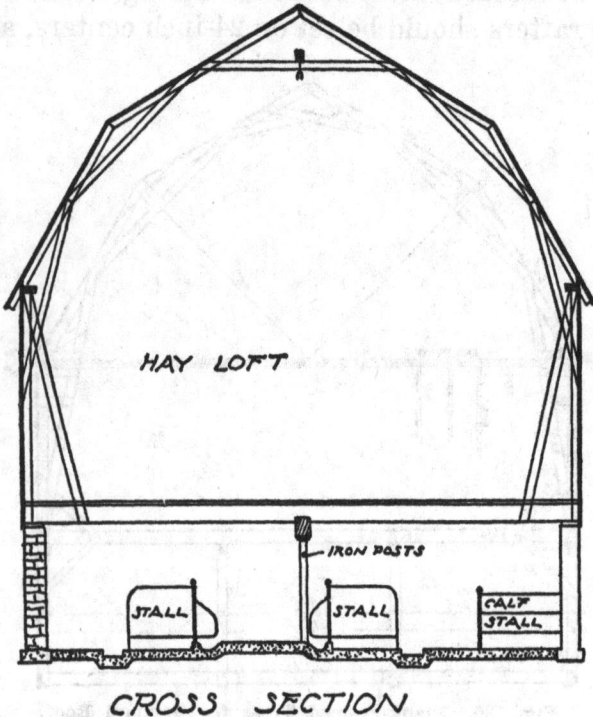

HAY LOFT

IRON POSTS

STALL   STALL   CALF STALL

CROSS SECTION

Fig. 137. Arrangement of Dairy Barn.

centers, and braced as shown in Fig. 136, can
be recommended. The rafters should set
directly over the studding, and be braced to
same. The floor-joist should be tied to each
other, either by letting them lap, or by nailing

a board on the side. All parts should be framed accurately, and well spiked.

Fig. 137 is a cross-section of a large dairy barn; it clearly shows the general arrangement of stalls, mangers, gutters, etc., all constructed out of cement laid on solid ground. The stall partitions are built up out of wrought-iron bars and pipes, leaving nothing to get out of order or decay. The wood superstructure is constructed

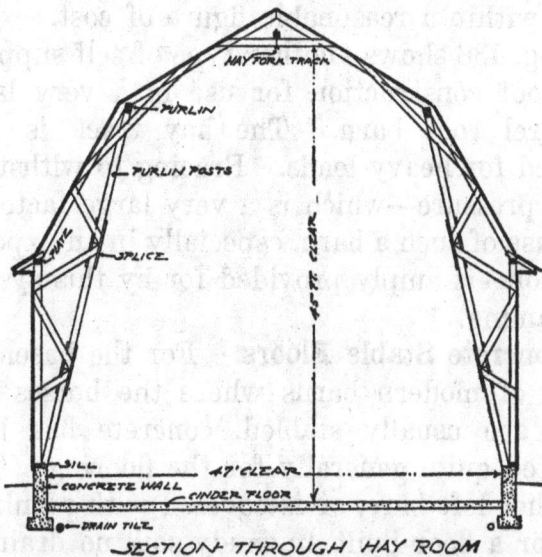

Fig. 138. Self-Supporting Roof Construction.

out of plank; and the roof is self-supporting, without posts or purlins, through each set of rafters being braced, forming a continuous arch from one sill to the other.

This roof gives an enormous capacity to the

hay room, and is well braced against sagging
and wind pressure.

The framing arrangement for the hay track
support is good, since it distributes the stress
evenly to all the roof bracing without an undue
amount coming on the top rafters.

The exterior of the barn is sided with
matched siding, and the roof is of shingles,
making a very durable and good-looking building, and at the same time a barn that can be
built within a reasonable figure of cost.

Fig. 138 shows another type of self-supporting roof construction for use in a very large
gambrel roof barn. The hay track is well
framed for heavy loads. Bracing to withstand
wind pressure—which is a very large factor in
the case of such a barn, especially in an exposed
location—is amply provided for by this system
of framing.

Concrete Stable Floors. For the basement
story of modern barns where the horses and
cows are usually stabled, concrete has been
adopted quite generally for the flooring. This
is either left bare, or is covered with plank.

For a floor built on sandy soil, no drainage
or foundation is necessary. If the soil is clay
or spongy, however, eight inches of coarse
gravel, stone, or cinders should be well tamped
and leveled before placing the concrete; but if
the soil is porous, then tamp the soil under the
center driveway.

When a foundation for floor is necessary,

drainage to conduct the water gathering in the foundation is also required; therefore the foundation must have large stone in the bottom to afford ample air space, and the excavation must have a slope to one or more points. These low points have tile drains leading the water from the building. This will insure dry floors, and prevent cracking from frost.

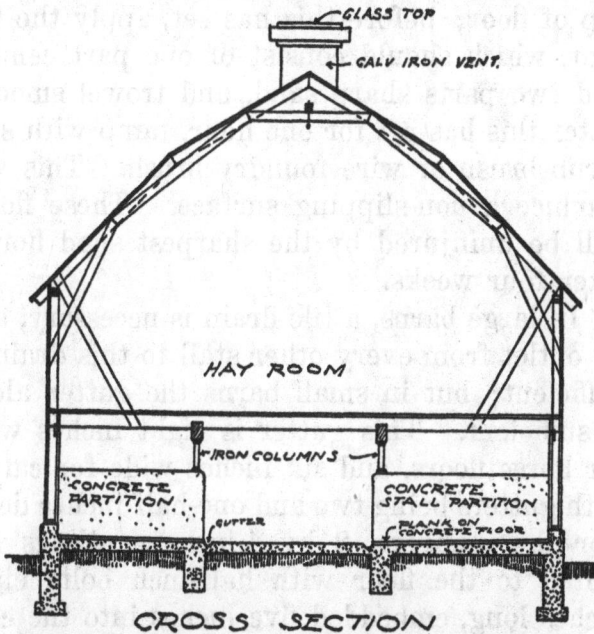

Fig. 139. Concrete Stable Floor and Partitions.

Four inches of concrete for horse and cattle floors is ample; and five inches for driveway will do, though some are made six inches thick. The proportion of concrete depends on the kind of

sand and aggregates. One of the following will
prove desirable: Cement one, sand two, gravel
three, makes very sound work. Cement one,
fine sharp sand one, coarse sharp sand two,
gravel or crushed stone four parts, will equal
in strength the first formula, and save much in
cost. Add sufficient water to make a stiff grout,
and tamp well into place so that the top will
be one-half to one-quarter inch lower than the
top of floor; before this has set, apply the top
coat, which should consist of one part cement
and two parts sharp sand, and trowel smooth.
After this has set for one hour, tamp with stiff
scrub brush or wire foundry brush. This will
produce a non-slipping surface. These floors
will be uninjured by the sharpest shod horses
after four weeks.

In large barns, a tile drain is necessary, and
an outlet from every other stall to this drain is
sufficient; but in small barns the gutter alone
is sufficient. This gutter is eight inches wide
for horse floors, and six inches wide for cattle,
both gutters being two and one-half inches deep.
For the mangers, 2 by 4-inch scantlings are
bolted to the floor with half-inch bolts eight
inches long, embedded five inches into the concrete. Manger partitions are boarded on both
sides and filled between with tamped concrete
four feet high, so that when the wood partition
has served its purpose a solid concrete wall will
remain. Use 4 by 4 timbers for framing stall
partitions (see Fig. 139).

Water-Tight Barn Floors. Sometimes it is desired to construct a floor in a horse barn to catch all fluids and keep them from going through into the basement. One way suggested is to lay a tight floor of 7/8-inch matching, then cover that first with asphalt about 1/2 to 3/4 inch, and on that lay a 1 3/4-inch matched floor. This floor should be properly graded, so as to drain to a trough for carrying off fluids.

This method is all right, but it would prove to be expensive in the course of a few years. Another and possibly a better way would be to lay a rough floor on the joists, and put on this floor 4 or 5 inches of good concrete, well laid down, not leaving more than 25 square feet in one block, using 1-inch expansion joints filled with pitch. This will keep the floor from cracking, and will also be water-tight. There should be a gutter just behind the horses, and the floor should have at least 1/2-inch fall to each foot. The floor should not be troweled smooth, but left rough, except in the gutter; put on this concrete (after it has been down six or seven days) about 2 inches of clay. Wet it thoroughly and tamp lightly into place. Clay is one of the best materials for horses to stand on.

If it seems preferable to have wood for the animals to stand on, lay a floor of rough planks, somewhat open, over the concrete, leaving cracks wide enough so that all liquid will immediately run through to the concrete and be drained to the desired points.

How to Make Barn Doors

Large-size barn doors that are strong and rigid, yet do not take up too much room in thickness, are the kind most wanted.

Fig. 140 shows two sectional drawings with elevations suitable for barn doors. The first is made of three thicknesses of boards as shown. The center is of 7⁄8-inch boards placed verti-

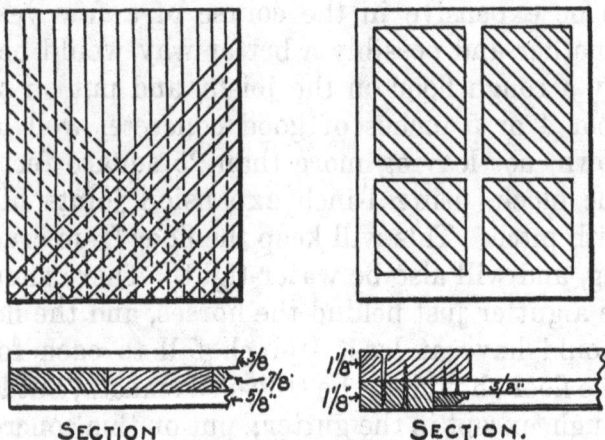

Fig. 140. How to Make a Heavy Barn Door.

cally, and 5⁄8-inch ceiling placed diagonally on both sides, covering the whole space, and well nailed. This will make a door 2⅛ inches thick.

The second is made of two 1⅛-inch pieces for the framework, lapped and screwed together. The panel work is made of 5⁄8-inch ceiling cut in and nailed with a stop-mould to cover the nailheads. All the laps and joints should be painted with white lead paint. This will make a door 2¼ inches thick.

Hay-Mow Doors. The best way to hang hay-mow doors for a barn with hay track high up in the gable, the door eight or ten feet wide, and made to swing clear back, is a point that often gives trouble to the barn builder who wants to make a neat job.

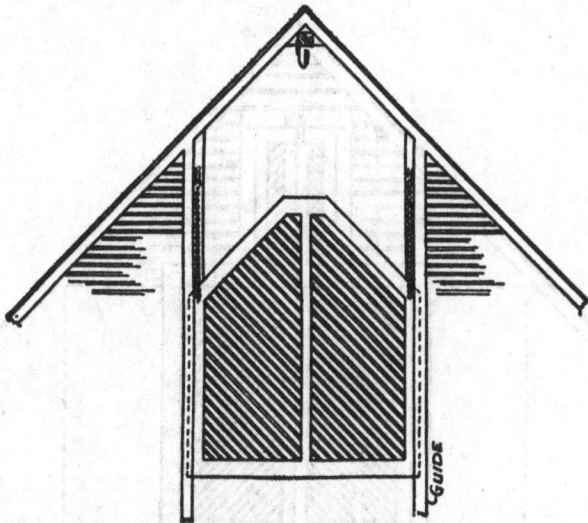

Fig. 141. Hay Door Hung with Weights.

Figs. 141 to 145 show the various arrangements. Fig. 141 is for a single door hung with weights and run in grooved jambs on the outside of the building. With this kind of arrangement, the door is made to slide up and down, and in this way can be made to slide close up to the comb.

Fig. 142 is for double doors, with small doors from these up to the carrier. With this arrangement, the doors can be hung with butts, and will swing clear of the cornice; but there should be a movable cross-bar at the top of the big doors to give a solid bearing to shut against. The bar can be removed when putting in hay, thus leaving a clear space down to the larger opening,

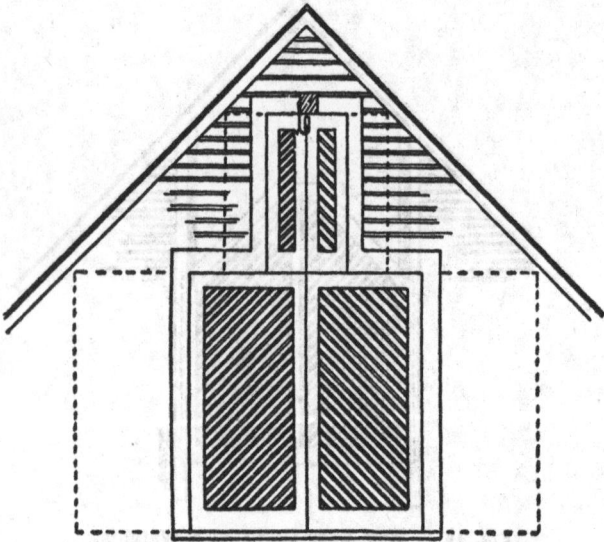

Fig. 142. Double Hay-Mow Doors.

as it is not particularly necessary to have the full-sized opening run all the way up to the carrier.

Fig. 143 is a sketch of **double hay doors** for a barn that are very satisfactory. The general instructions for making are as follows:

Lay off the door in halves to fill the opening clear up to the track. Next begin at the outer upper corner and lay a downward line towards the center of the door at the same angle as the pitch of the roof. Cleat and saw on this line, and put hinges on the inside. This top **leaf** will

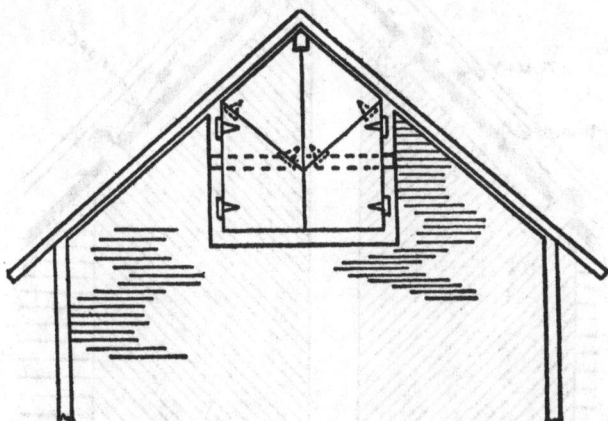

Fig. 143. Double Doors to Swing and Fold Down.

then fold down as the door opens, and all will swing under the cornice and lie flat against the wall.

To fasten the doors open, make a double-ended spiked pole, with which press the outer or top fold of the door up firmly against the fascia, seating the lower end of spiked pole near the bottom of the door. There should be a 2 by 4 or a 2 by 6 bar across the opening to hook the doors to, as well as for safety, while opening or closing.

Fig. 144 is a sketch of another way of hanging large barn doors, which has given good satisfaction wherever it has been used.

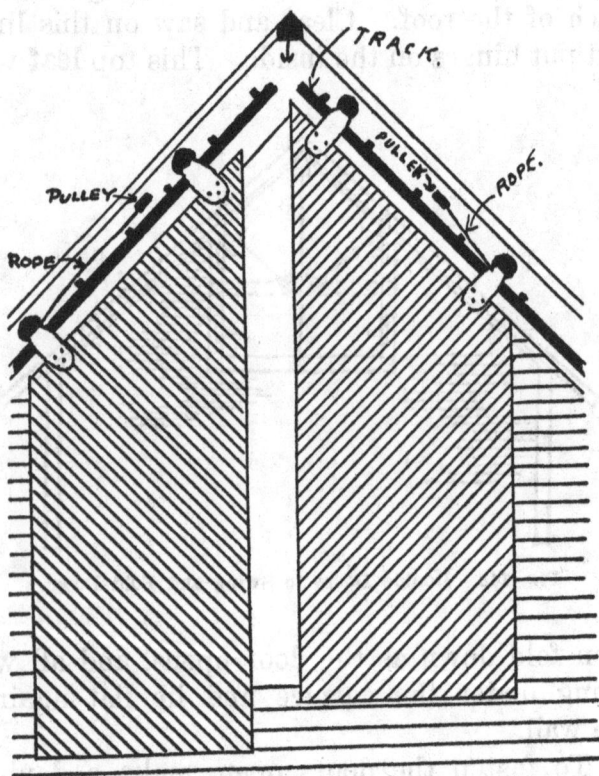

Fig. 144. Weighted Doors on Inclined Tracks.

Any flat barn-door track will do. A ⅝-inch rope is tied to the lower hanger and run through a 5-inch wooden pulley set in rafter; from there it is run down inside of rafter 3 or 4 feet to another pulley; and from there to a small weight below, which will run between two studs. Hanging a door this way does not take as heavy a weight as running it up and down the grooves, and it runs more easily. When doors are both closed, they can be fastened together on the inside with a hook. Bumpers on the outside of the barn prevent them from running off the track.

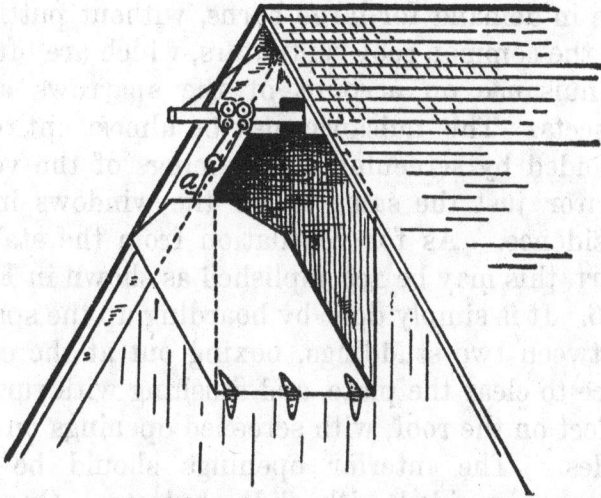

Fig. 145. Hay-Mow Door to Fold Down.

A practical barn builder writes concerning hay-mow doors as follows:

"For some time I thought to hang the door with weights was the only way; but the last one of that kind I put up, two of us worked over an hour trying to get it to work right; finally we made the guides so loose that the door would nearly fall out. It did not work well then. If the wall springs either in or out, the door will bind.

"The enclosed illustration (Fig. 145) will give you my plan. The wire, **a,** has a hook on the upper end to hook into the pulley, as shown. A pull on the hay-rope will raise the door with ease. The carrier trips the same as with a load of hay. The door is pulled up close, or left open a little for ventilation. Tie the hay-rope any place, and it holds the door, which, being hinged at the bottom, hangs below the opening, out of the way."

Ventilating a Barn. Systems of ventilation are in demand for large barns, without putting on the common roof ventilators, which are often a nuisance on account of the sparrows and insects. This nuisance may be almost entirely avoided by screening the openings of the ventilator just the same as for the windows in a residence. As for ventilation from the stable part, this may be accomplished as shown in Fig. 146. It is simply done by boarding up the space between two studdings, boxing out at the cornice to clear the plate, and finishing with turret effect on the roof, with screened openings on all sides. The interior openings should be as shown, provided with slide shutters. One of these vent shafts should be placed about every eight feet, or opposite every other stall.

Barn Windows. As barn windows, as a general rule, are one-sash windows, it is sometimes difficult to contrive a convenient way to open them. The slide window is not well liked, because it is almost impossible to make it so that rain will not beat in. Besides, it is necessary to have a stop outside to guide the sash, which at best is a dirt catcher and holds moisture, causing the window-sill to rot in a short time.

Fig. 147 illustrates, at the right, a type of basement window which has given good satisfaction. It being impossible to raise the basement windows on account of the main sill just above them, the frames are made as shown. This allows the window to be opened by drawing top of sash back onto the pieces at-

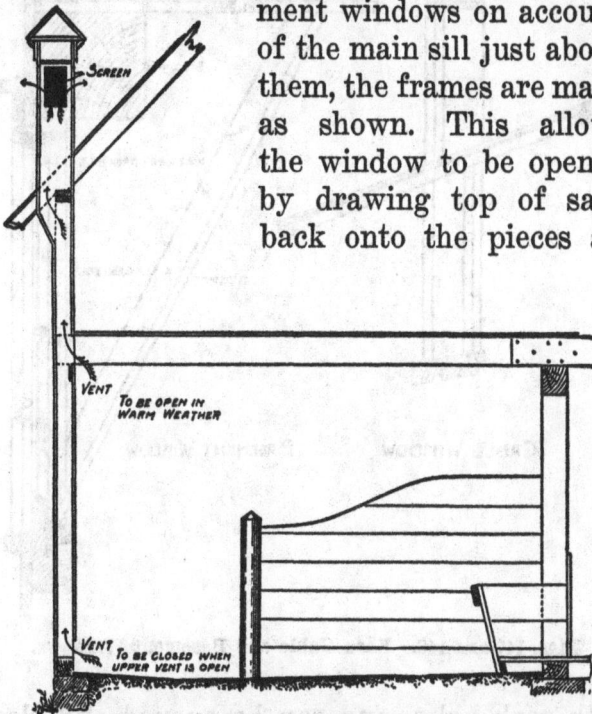

Fig. 146. How to Ventilate a Stable.

tached to frame and floor above, which will give ventilation without a direct draught; or sash can be raised as far as floor will allow. Sometimes the side jambs are made wide at the top, so that there is no opening at the sides, when

the window is open; but this is not necessary
unless the window is close to the live-stock.

A gable window is shown also, in Fig. 147,
at the left. It is hung just above the center

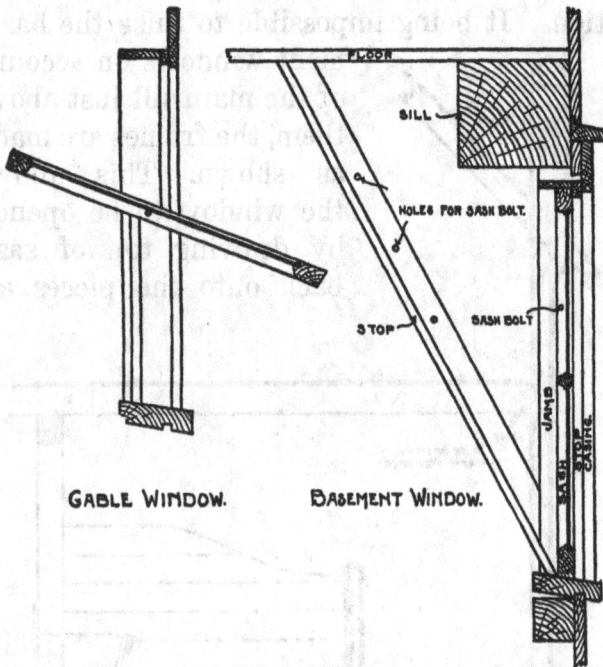

FLOOR

SILL

HOLES FOR SASH BOLT.

STOP

SASH BOLT

JAMB

SASH

STOP

JAMB

GABLE WINDOW.   BASEMENT WINDOW.

Figs. 147 and 148. Barn Gable and Basement Windows.

with sash bolts, and can be opened and closed
with cords running to some convenient place,
as not everybody likes to climb to the peak of
a high barn when it is empty, to open or close
windows.

Framing of Factories, Stores and Public Buildings

When we come to the framing of larger buildings, such as factories, stores, halls, etc., the work becomes more complicated and technical in its nature, falling perhaps more within the province of the structural engineer than the ordinary carpenter and builder. Nevertheless, certain of the fundamental principles of such work should be understood, and may well be considered here. They are, briefly: **standard mill construction; strength of timbers; principles of truss construction;** and **framing of public buildings** for safety and also for architectural effect.

STANDARD MILL CONSTRUCTION

The succession of heavy fire losses each year is the penalty which this country is paying for the erection of light, cheap, and poorly designed buildings. The cost of fire insurance is a direct yearly tax on the building and its occupants. It is the duty of those responsible for the design of buildings, to plan them so that this tax may be the smallest possible, and this can be done often without any great increase in the cost of the building itself.

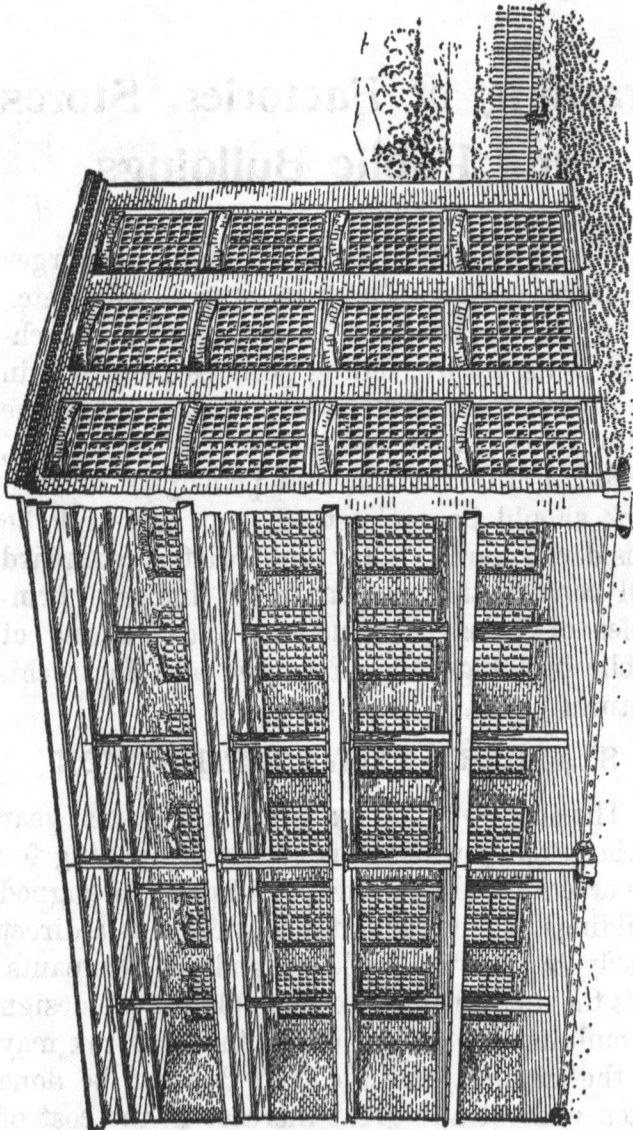

Fig. 149. Standard Mill Construction—General Framing Plan.

According to tests made by the Boston Insurance Engineering Experiment Station—which only confirm and bear out the experience of years—it has been clearly shown that, all things considered, the **mill** or **slow-burning** type of construction is to be recommended for most factory and warehouse buildings. In some cases, where the contents will be extra inflammable, the extra expense of the thoroughly fireproof reinforced concrete structure is warranted; otherwise mill construction should be used.

In order that there shall be no misunderstanding of what is meant by **mill construction,** we shall say that it consists in disposing the timber and plank in heavy solid masses so as to expose the least number of corners or ignitable projections to fire. Also it consists in separating every floor from every other floor by fire stops.

The essential features of standard mill construction are illustrated in Fig. 149, and are, briefly, as follows:

1. The walls should be of brick or concrete block, at least 1 foot thick (16 inches for best work) in top story, and increased in thickness at lower floors to support additional load. The pilastered wall has many advantages, and is often preferred to the plain wall. Window and door arches should be of brick; window and outside door-sills and underpinning, of granite or concrete.

2. The roofs should be of 3-inch pine plank,

spiked directly to the heavy roof timbers and
covered with 5-ply tar and gravel roofing.
Roofs should pitch ½ inch to ¾ inch per foot.
An incombustible cornice is recommended when
there is exposure from neighboring buildings.

3. Floors are best made of spruce plank 4
inches or more in thickness according to the
floor loads, spiked directly to the floor timbers

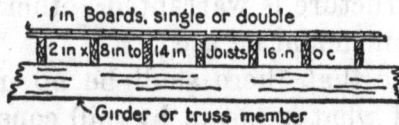

Fig. 150. Undesirable Floor Construction.

and kept at least ½ inch clear of the face of the
brick walls. Figs. 150 and 151 show bad and
good forms of floor construction. In floors and
roof, the bays should be 8 to 10½ feet wide; and
all plank two bays in length, laid to break joints

Fig. 151. Good Types of Floor Construction.

every 4 feet, and grooved for hardwood splines.
Usually top floor of birch or maple is laid at
right angles to the planking; but the best mills
have a double top floor, the lower one of soft
wood laid diagonally upon the plank, and the

upper one laid lengthwise. This latter method
allows boards in alleys to be easily replaced
when worn, and the diagonal boards brace the
floors, reduce vibration, and distribute the floor
load even better than the former method.

Between the planking and the top floor
should be two or three layers of heavy tarred
paper, laid to break joints, and each mopped
with hot tar or similar material to produce a
reasonably water-tight as well as dust-tight
floor.

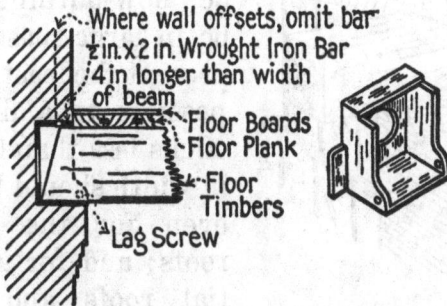

Fig. 152. Cast-Iron Wall Box for Floor Timbers, with Lugs for
Anchoring to Walls.

Rapid decay of basement or lower floors of
mills makes it desirable, whenever wood is not
absolutely necessary, to provide cement floors
for these places. If wooden floors are required,
crushed stone, cinders, or furnace slag should be
spread evenly over the surface and covered with
a thick layer of hot tar concrete, on which is
often laid tarred felt, well mopped with hot tar
or asphalt. On this a floor of 2-inch seasoned

plank should be pressed, nailed on edge without perforating the waterproofing under it, and the hardwood top floor boards nailed across the plank. Cement concretes promote decay of wood in contact with them. If extra support is required for heavy machinery, independent foundations of masonry should be provided.

4. In regard to timbers and columns it should be remembered that all woodwork in standard construction, in order to be slow-burning, must be in large masses that present the least surface possible to a fire. No sticks less than 6 inches in width should be used, even for the lightest roofs; and for substantial roofs and floors, much wider ones are needed. Timbers should be of sound Georgia

Fig. 153. Floor Timber Resting on Cast-Iron Wall-Plates, with Lugs for Anchoring Timber to Wall.

pine; and for sizes up to 14 by 16 inches, single sticks are preferred; but timbers 7 or 8 inches by 16 are often used in pairs, bolted together without air-space between. They should not be painted, varnished, or filled for three years because of danger of dry rot; and an air-space should be left in the masonry around the ends for the same reason. Timbers should rest on cast-iron plates or beam boxes, in the walls, and on cast-iron caps on the columns, as shown in Figs. 152, 153, 154, and 155.

Beam boxes are of value, as they strengthen the walls when floor loads are heavy and distance between windows small; they facilitate the laying of the brick and the handling of the beams; and there is less possibility of breaking away the brick in putting the beams in place.

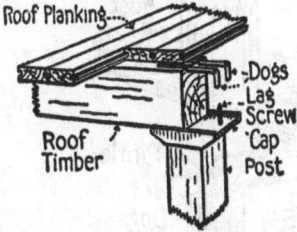

Fig. 154. Roof Timber Resting on Column Cap—Timbers Held Together by 1-Inch Wrought-Iron Dogs.

They also insure a proper air-space around beams.

Columns should be set on pintles, which may be cast in one piece with the cap, or separately, as preferred (see Figs. 156 and 157.) Columns of cast-iron are preferred by s o m e engineers; and, w h e n t h e building is equipped with automatic sprinklers, h a v e proved satisfactory, but are not so fire-resisting as timber.

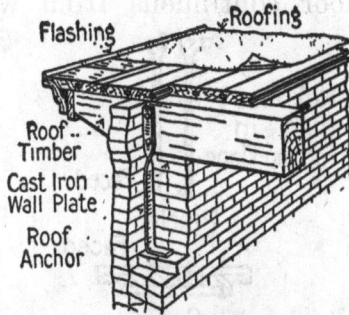

Fig. 155. Roof Timber Resting on Cast-Iron Wall-Plate, Showing Overhanging Open Wood Cornice and Wrought-Iron Anchor

Wrought-iron or steel col-

umns should not be used unless encased with at
least 2 inches of fireproofing.

One of the most important features of slow-

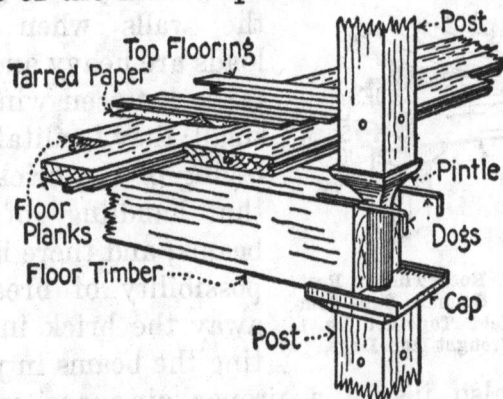

Fig. 156. Cast-Iron Cap and Pintle for Columns and Dogs for
Holding Floor Timbers Together.

burning construction is to make each and every
floor continuous from wall to wall, avoiding

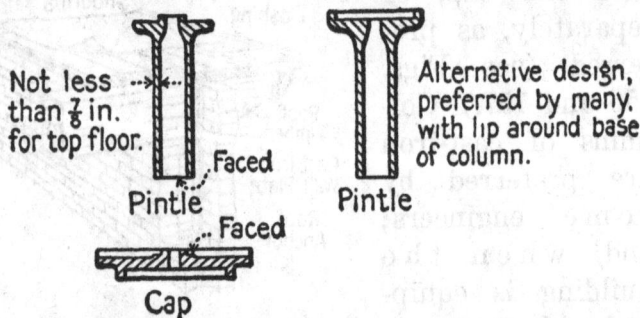

Fig. 157. Cap and Pintle Cast to Fit Columns.

holes for belts, stairways, or elevators, to the
utmost extent possible, so that a fire may be
confined to the floor where it starts. No well-
informed mill owner, engineer, or builder will

therefore fail to locate elevators, stairs, as well as main belts, in brick towers or in sections of a building cut off by incombustible walls from all the rooms of a factory. Openings in these walls should be provided with fire-doors, preferably self-closing. These should be hung on heavy, inclined, solid steel rails at least $3\frac{1}{2}$ inches by $\frac{3}{8}$ inch, and balanced by a weight held by a fusible link.

In modern practice all belts and ropes which may be used for transmission of power to the various rooms are placed in incombustible vertical belt-chambers, from which the power is transmitted by shafts through the walls into the several rooms of the factory. There should be no unprotected or unguarded openings in the inner walls of this belt-chamber.

Saw-Tooth Roof Construction. The great advantages and the increasing use of **saw-tooth** roof construction, together with the lack of familiarity with it in many sections, make it desirable to outline its important features.

Two typical designs are illustrated—one, Fig. 158, a textile weave shed with good basement for shafting for driving looms on main floor above, thus dispensing with the overhead shafting and belting in the weave room; the other, Fig. 159, a design for a light machine shop or foundry. Other designs are applicable, with light wooden trusses or reinforced concrete.

The important advantages of this form of roof construction are:

Uniform diffusion of light throughout the room, thus making all space in it available. With all interior surfaces painted white, and with ribbed glass in the sash, the diffusion of light is almost perfect.

Adaptability for lighting large floor areas in wide buildings with low head room, compared with what is necessary in wide buildings with the ordinary form of monitor skylights.

They provide the true solution to the problem of excluding the direct rays of the sun and obtaining the very desirable north light in all sections.

Economy in lighting, in that they lessen the fixed charges due to the lessened number of hours per day during which artificial light is necessary.

Better working conditions, especially in textile mills, therefore increasing production and encouraging permanency of the help.

The saw-tooth form is especially adapted to buildings for weaving and similar processes in textile factories, machine shops, foundries doing light work, and for similar or allied operations, such as assembling and drafting; and in some dye houses, where careful matching of colors is necessary.

As to the disadvantages, while testimony of those having had experience with saw-tooth roofs is almost uniformly favorable, more or less difficulty has been experienced. Practically all of it, however, may be summed up as due

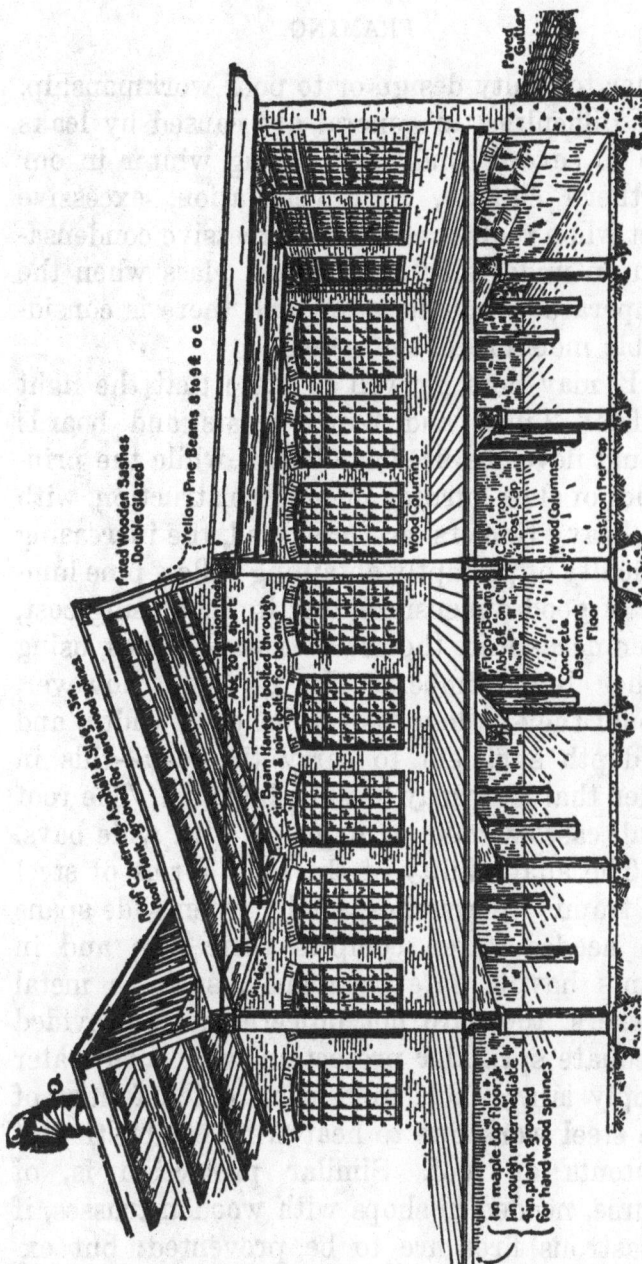

Fig. 158. Textile Weave Shed—Standard Mill Construction with Saw-Tooth Roof.

either to faulty design or to poor workmanship. The difficulties in general are caused by leaks due to severe conditions during winter in our northern climate, poor ventilation, excessive heat when roofs are thin, or excessive condensation on under side of roof and glass when the temperature outside is low and there is considerable moisture in the rooms.

It may here be well to state that the light roof of 2-inch and 3-inch joists and boards should never be used; and that, while the principles of slow-burning or mill construction, with the heavy timbers, are preferred, the increasing difficulty of promptly obtaining yellow pine lumber of good dimensions, and its increasing cost, often necessitate the use of trussed forms, using rather light timbers. In no case, however, should they be less than 6 inches in width, and of depth sufficient to carry the load—this in order that they may be slow-burning. The roof in all cases should be of plank, with wide bays.

The adaptability of the light forms of steel for framing trusses, especially when wide spans are needed, often compels their use; and in plants having safe occupancy, such as metal workers, they are not objectionable, provided adequate sprinkler protection with good water supply is available to prevent quick failure of the steel work due to heat from combustion of contents or roof. Similar protection is, of course, needed in shops with wooden trusses, if disastrous fires are to be prevented; but ex-

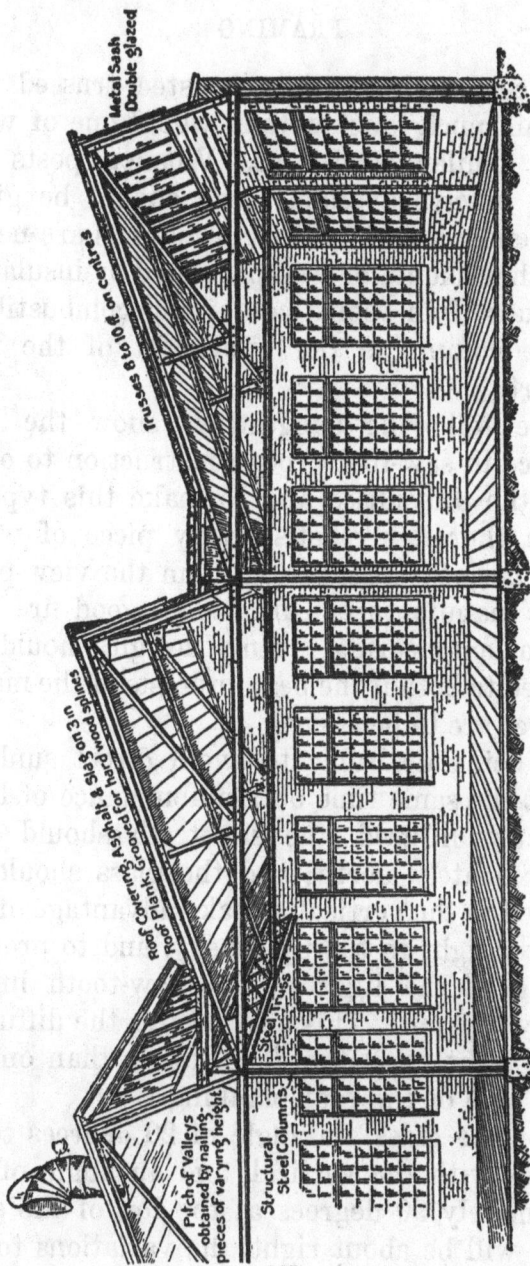

Fig. 159. Light Machine Shop—Saw-Tooth Roof Construction.

Metal Sash Double glazed

Trusses 6 to 10 ft. on centres

Roof Covering, Asphalt & Slag on 3 in.
Roof Plank. Grooved for hard wood splines

Pitch of Valleys
obtained by nailing
pieces of varying height

Structural
Steel Columns

Steel Trusses

perience has shown that the steel-trussed roof will fail much quicker than would one of wood under similar conditions. Wooden posts are nearly always available, and should be given preference; but if light steel columns are necessary, they should be well protected by insulating materials if in rooms containing combustibles, as the column is the vital part of the roof support.

The following suggestions show the best practice in saw-tooth roof construction to overcome the difficulties and to make this type of roof a thoroughly satisfactory piece of work. What is good engineering from the view-point of the manufacturer can also be good fire protection engineering. Any design should be adapted to both if the best interests of the manufacturer are to be served.

It being desirable to avoid direct sunlight and at the same time obtain abundance of light perfectly diffused, the saw-teeth should face approximately north; and the glass should be inclined to the vertical, to take advantage of the brighter light in the upper sky, and to prevent cutting of the light by the saw-tooth immediately in front. This also assures the diffusion of the light upon the floor rather than on the under side of the roof planking.

For the glass, an angle of 20 degrees to 25 degrees with the vertical, and an angle of approximately 90 degrees at the top of the saw-tooth, will be about right, the variations to depend on the amount of light required, and on the latitude. A sharper angle at the top is not needed, as it increases the cost, there being more roof to cover and larger spans. More glass is also required in proportion, and the light is not so good, more sky light being lost and too much thrown on under side of roof.

Double glazing, with space between, is preferred on account of its conducting qualities; but is not always necessary, except in the North.

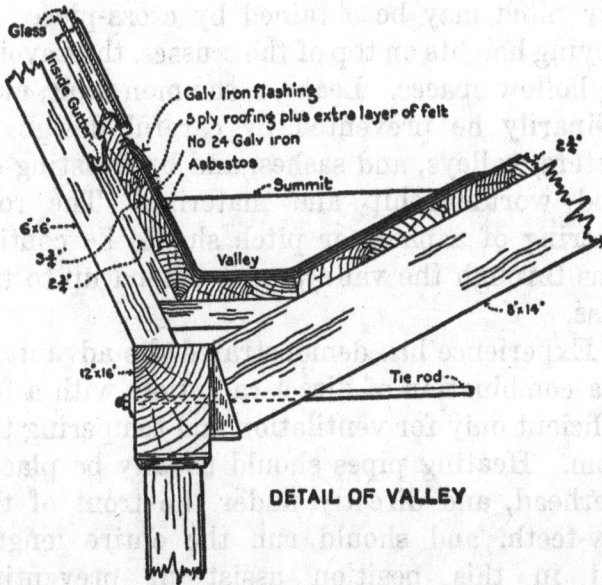

Fig. 160. Saw-Tooth Roof Framing—Detail of Valleys.

The inside glazing should be factory ribbed glass, with ribs vertical and inside. Shadows cast by trusses are then almost unnoticeable.

Condensation gutters, as shown in the detail drawing, Fig. 160, are needed inside at the bottom of the sash; and they should be drained through inside conductors, and not outside, under bottom of sash, as these latter admit cold air and are liable to freeze.

Valleys between the saw-teeth should be flat, 14 inches to 2 feet in width; and should pitch one-half inch per foot toward the conductors, which should be of ample size, and not much over 50 feet apart, preferably less. The necessary pitch may be obtained by cross-pieces of varying heights on top of the trusses, thus avoiding hollow spaces. Leaks, a common fault, may ordinarily be prevented by careful design of gutters, valleys, and sashes, and by insisting on good workmanship and materials. The roof covering of asphalt or pitch should be continuous through the valleys, and extend up to the glass.

Experience has demonstrated the advantage of a combination of direct radiation, with a fan sufficient only for ventilation and tempering the room. Heating pipes should usually be placed overhead, and directly under the front of the saw-teeth, and should run the entire length, and in this position assist in preventing condensation.

Where there is no moving shafting, some forced circulation is necessary. This is best obtained by a fan, often driving air from a dry basement or outside as required, and discharging it over heating coils to the floor above. In weave and similar rooms, is this especially necessary and advantageous in promoting health and comfort of employees, making greater efficiency possible.

Ventilation and cooling of these large areas with comparatively low stories must not be neglected. Ample vents are needed at top, in shape of large metal ventilators with double walls and tight dampers. They are recommended instead of pivoted or swinging sash, which are apt to leak in driving storms, and which, when open, allow dirt to blow in off the roof. Good windows are advised in side walls, and experience has shown their value.

Framing of the saw-teeth may be in timber, steel, or reinforced concrete. The design should be such as to obstruct the light as little as possible; strong enough to hold wet snow without sagging; and stiff enough to carry shafting motors, etc., when they are to be overhead. When wood or steel is used, the roof planking should be 3 inches or over, spanning wide bays of 8 to 10 feet.

Hollow spaces in roofs should not be permitted. They are very undesirable from a fire standpoint, and any condensation which may take place in them during cold weather soon rots both plank and sheathing.

Sheathing, even without spaces behind it, is more or less a bad feature, as it is readily combustible; but, if used, should be applied directly

to the under side of the roof plank, with only a
layer of some insulating material between, so
that there may be no concealed space. If 3-inch
plank is sufficient for a flat roof, it should be for
a saw-tooth; and, with good circulation of air,
there should be no trouble, except in wet rooms,
where condensation is bound to occur whether
under a roof or the floor of the room above, unless large quantities of dry air are discharged
into the room.

Saw-tooth roofs necessarily cost more, as
there is practically the same amount of roofing
as in flat roofs, and in addition there is the cost
of windows, glazing, flashing, conductors, condensation gutters for the skylights, and a somewhat larger cost of heating. The additional cost of these items does not, however,
fairly represent comparative cost, as there
should be considered the total cost of the building compared with that of an ordinary one of
sufficiently high stories and narrow enough to
give the required light. When this is done, the
slight additional cost is far outweighed by advantages of the saw-tooth type for work where
good light is desirable.

Strength of Timbers

Strength of Beams Supported at Both Ends.
"Will you decide a little argument which my mate and I have been having?" said a carpenter friend of mine as he and his mate came into the office the other morning. "We want to know which is the stronger of two pitch pine beams we have outside here. One is 9 inches by 6 inches, and the other 8 inches by 7 inches; and both are to be used over openings of 12-feet span. My mate thinks the 9 by 6 is the stronger of the two, but I hold that the 8 by 7 will carry more weight."

"I shall be very pleased to work it out for you, and can tell you the result in about a minute;" said I, "but if you have half an hour to spare, I should prefer to show you how I arrive at my figures, and thus enable you to make the necessary calculations for yourselves whenever you desire."

As work was not very pressing that morning, they readily agreed to take a lesson, and I proceeded somewhat as follows:

A piece of wood 1 inch square placed on bearings one foot apart, will break under a certain weight. This wieght varies with different woods and with different specimens of the same wood; but most authorities have agreed to regard certain average weights as standards. These averages were obtained from hundreds of experiments, and are therefore fairly reliable. The following table deals with a few woods only, but is sufficient for our present purpose.

Breaking weights of wood beams 1 foot long, 1 inch broad, and 1 inch deep, loaded in the center and supported at both ends (the length means the span of the opening—that is, the distance in clear of bearings), are as follows:

Ash ....7 cwt.
White oak ....$5\frac{1}{2}$ "
Georgia pitch pine..................5 "
Norway red pine.....................4 "
Spruce ....$3\frac{1}{2}$ "
Teak ....8 "
(1 cwt.=112 pounds.)

Referring to the illustration, Fig. 161, we find at **A**, that, taking pitch pine as our wood, a piece 1 inch by 1 inch, on bearings 1 foot apart, will break with a central load of 5-cwt. Now, it is quite clear that if we increase the breadth to 3 inches, as at **B**, it will take three times five, or 15 cwt., to break the piece.

But suppose that we put this piece of 3 by 1 pitch pine on edge and see what it takes to break it. Instead of 15 cwt., we shall find that it takes no less than 45 cwt. to do so. Or, as the books put it, **the strength of a beam is as the square of its depth.** That means that instead of saying, as in **B**, three times five, we square the three, and say three times 3 are nine, and nine times

5 cwt. is 45 cwt., which is the breaking weight (approximately) of a piece of 1 by 3 pitch pine on edge over a 12-inch bearing.

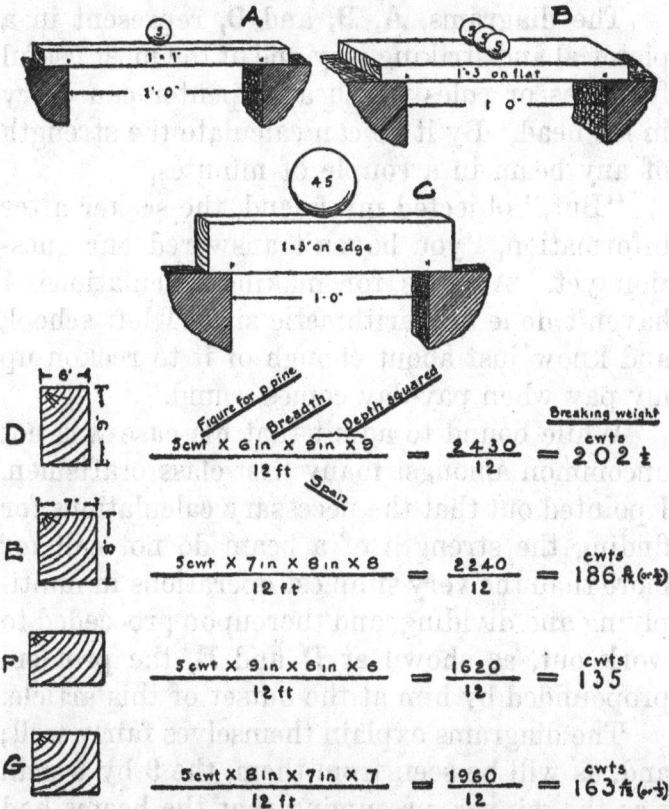

Fig. 161. Ultimate Strength of Beams Supported at Both Ends with Load Concentrated at the Middle.

But suppose, further, that instead of the bearings being 12 inches apart the distance between them had been 2 feet. It is clear that the beam would only carry half as much, and we should have to divide our answer by two. And as the longer the beam, the less it will bear, this gives us another rule which will be referred to later.

The diagrams, **A**, **B**, and **C**, represent in a pictorial and striking way one of the most useful formulas or rules which a carpenter can carry in his head. By it he can calculate the strength of any beam in a couple of minutes.

"But," objected my friend, the seeker after information, "you haven't answered our question yet. And, as for making calculations, I haven't done any arithmetic since I left school, and know just about enough of it to reckon up my pay when pay-day comes round."

While bound to admit that his case was not uncommon amongst many first-class craftsmen, I pointed out that the necessary calculations for finding the strength of a beam do not call for more than the very simplest operations in multiplying and dividing, and thereupon proceeded to work out, as shown at **D** and **E**, the problem propounded by him at the outset of this article.

The diagrams explain themselves fairly well; and, as will be seen from them, the 9 by 6 man was the winner, presuming that the beams had been placed (as they should be) with their greatest dimensions upright.

But suppose the same beams had been laid flat instead of on edge, how would they work out then? **F** and **G** give the results, which show that the 8 by 7 beam is a good deal stronger than the 9 by 6 if used in this position.

It must be observed, of course, that these are the breaking weights for beams, and it is obvious that the safe load is what is wanted. This is found by dividing the breaking weight by a certain factor (known as the **factor of safety**)— generally 4 or 5 for a dead load, and 8 or 10 for a live load.

It must also be remembered that the foregoing formula is for a beam loaded in the center. **If the load is distributed evenly throughout the length of the beam, it will carry just double what it would if loaded in the center.**

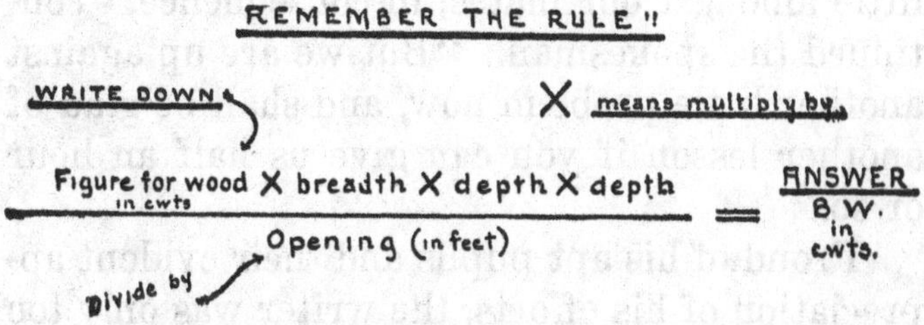

REMEMBER THE RULE !!

WRITE DOWN

X means multiply by.

$$\frac{\text{Figure for wood} \times \text{breadth} \times \text{depth} \times \text{depth}}{\text{Opening (in feet)}} = \frac{\text{ANSWER}}{\text{B.W. in cwts.}}$$

Divide by

Fig. 162. To Find Breaking Weight for Beams Supported at Both Ends—Load Concentrated at the Middle.

To sum up the matter, the points to carry in one's head are these: First, the figure given in the table for the particular wood. Second, the way of putting down the simple sum or formula, which must be as shown in Fig. 162. Third, divide your result by figure (factor) of safety, say five for dead load and ten for live load. (This gives safe load for center of beam. If load is distributed, multiply this by two.)

Thus divested of all formal language, this useful little working formula was easily grasped by my two interrogators; and, in the hope that it may prove similarly useful to many others, the writer begs to present it.

Strength of Beams Supported at One End. "More work for the calculator," was the remark which greeted the writer the other day, as his two carpenter friends once more came into the office. "We have been applying your last little lesson on the strength of beams to several cases which have occurred in our work lately, and have found no difficulty in arriving at correct results. In fact, we have been 'showing off' a little amongst our mates, in consequence," continued the spokesman. "But we are up against another little problem now, and shall be glad of another lesson if you can give us half an hour or so."

Proud of his apt pupils and their evident appreciation of his efforts, the writer was only too pleased to put his services again at their disposal, and, after a brief talk, found the problem to be as follows:

A wooden beam was to be fixed so as to project some five feet from the face of a building, for the purpose of hoisting goods from the street level to a warehouse on the upper floor. (The technical term for a beam in this position is **cantilever**, and such a beam will be referred to by that name throughout this discussion.) A piece of pitch pine 7 inches by 5 inches had been selected for the job, and the question arose as to the amount of weight which could safely be hoisted upon it.

First of all, we ran over our last lesson on the strength of a beam when supported at both ends, and found out what load a piece of pitch pine, 7 inches by 5 inches, and 5 feet long, would carry if placed on edge, when supported at both ends and carrying a central load. Our rule, or formula, used in the last lesson was of course required.

This gave us the result shown in Fig. 163, at **A**—namely, 245 cwt. as the breaking weight required. It will be remembered that in the previous article it was stated that a beam with a distributed load will carry twice as much as the same beam with a central load, and that therefore the answer to this sum could be doubled for such conditions. But it is obvious that the strength of a beam supported at both ends is much greater than that of one supported (or fixed) at one end only, and is, relatively, as 4 is to 1.

Therefore, to find the breaking weight of our piece of 5 by 7 when fixed as a cantilever, we divided our result by 4, giving us 245÷4=61¼ cwts. But it was the amount our cantilever would carry which we wanted to find; and that brought in the question of the relation between the breaking weight and safe load—or, as it is termed, the **factor of safety**—to which reference was made before.

In this connection, the nature of the load or stress to which the beam is to be subjected is important, and may cause the safe load to vary

THE STRENGTH OF BEAMS

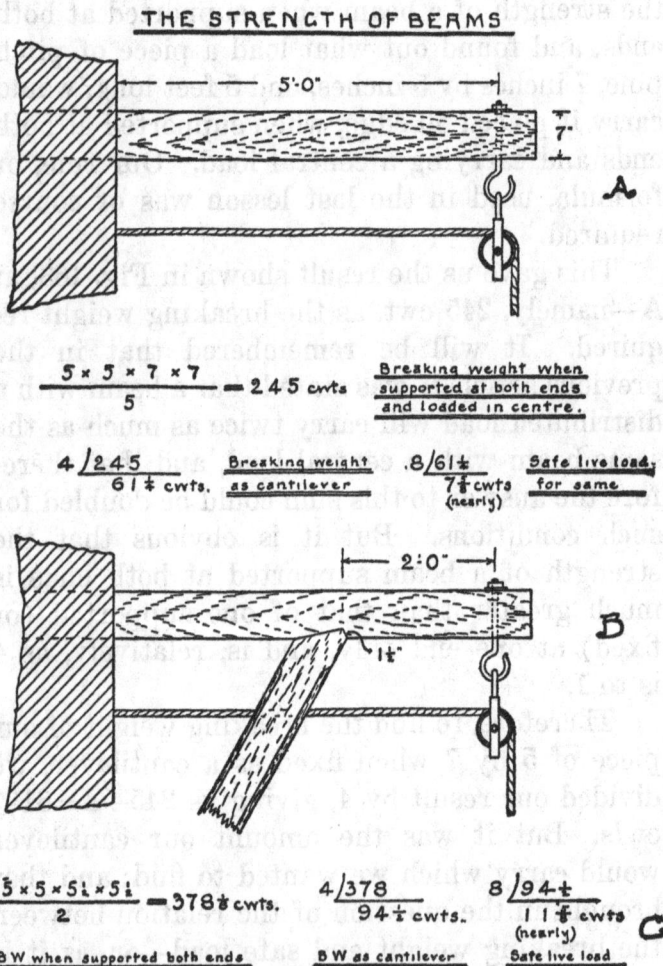

$$\frac{5 \times 5 \times 7 \times 7}{5} = 245 \text{ cwts}$$

Breaking weight when supported at both ends and loaded in centre.

$$4\,\underline{)245}$$
$$61\tfrac{1}{4}\text{ cwts.}$$

Breaking weight as cantilever

$$.8\,\underline{)61\tfrac{1}{4}}$$
$$7\tfrac{1}{2}\text{ cwts}$$
(nearly)

Safe live load, for same

$$\frac{5 \times 5 \times 5\tfrac{1}{2} \times 5\tfrac{1}{2}}{2} = 378\tfrac{1}{2}\text{ cwts.}$$

$$4\,\underline{)378}$$
$$94\tfrac{1}{2}\text{ cwts.}$$

$$8\,\underline{)94\tfrac{1}{2}}$$
$$11\tfrac{3}{4}\text{ cwts}$$
(nearly)

B W when supported both ends   B W as cantilever   Safe live load

Fig. 163. Ultimate Strength of Beams Used as Braced and as Unbraced Cantilevers. Loaded at Outer End.

from one-fifth to one-tenth of the breaking
weight, according as the load is a live or dead
one. Our load in this case was, of course, a live
one; but another consideration entered into the
question—namely, the manner of applying the
hoisting force. That is to say, we had to consider whether the force was to be applied in a
series of jerks such as given by sailors in pull-
ing on a block and tackle, or to be steadily and
continuously applied as by a winch or hoisting
drum. If in the first manner, the greatest margin of safety would have to be allowed; and, at
most, only one-tenth of the breaking weight
should be carried. For the continuous, steady,
pull, however, a factor of safety of one-eighth
would probably be sufficient. As the power in
this case was a drum driven by an electric
motor, we decided upon eight as our factor, and
applied it to our breaking weight ($61\frac{1}{4} \div 8$), obtaining approximately $7\frac{5}{8}$ cwts.

As this was a much greater weight than was
generally to be hoisted to the warehouse in question, my two carpenter friends felt quite satisfied when our calculations showed that their
guess at the size of the cantilever had been on
the safe side, and that it was strong enough for
any emergency.

A Braced Cantilever Beam. On inquiring as
to the reason why the point of suspension for the
load was so far out, it was explained that bulky
packages were carried up and down, and it was
therefore necessary that they should swing well

clear of the stories below. That led to the sug-
gestion that a strut or brace could be used under
the cantilever; and, as this would increase its
strength very materially, it was decided to calculate just what would be gained by so doing.

Taking the sketch of the strutted cantilever
(Fig. 163, **B**), we found that the amount overhanging the end of the strut was 2 feet. But the
under side had been weakened at the joint where
the strut had been let into the cantilever, and
the effective depth of the beam was thereby reduced by $1\frac{1}{2}$ inches, the depth of the shoulder.
Our sum, therefore, was changed, reading as at
C; and gave us for result $378\frac{1}{8}$ cwts. as the
breaking weight of a pitch pine beam 2 feet long,
$5\frac{1}{2}$ by 5 inches. Dividing this by 4 (neglecting
the odd $\frac{1}{8}$ cwt.), we obtained our breaking
weight for the cantilever $378\div4=94\frac{1}{2}$ cwts.
Again dividing this by 8, we obtained as our safe
load $94\frac{1}{2}\div8=11\frac{5}{8}$ cwts., approximately, thus
showing a considerable increase of strength over
the cantilever without the strut.

The foregoing example of a cantilever used
for permanent hoisting purposes is not, perhaps,
met with so commonly to-day as formerly, elevators inside of buildings having largely superseded them. For purposes of temporary work,
however, such as the raising of some heavy
article to the upper floor of a building, the piece
of timber projecting from some opening is still
frequently in evidence. The nearest carpenter
is generally called upon to rig up the affair, and

the formula worked out here (see Fig. 162) is an exceedingly useful one to have at hand in such cases. For, unlike beams, girders, joists, etc., the sizes of which are calculated in most cases by the designer of the building, these temporary rigs are left wholly to the skill and ingenuity of the carpenter, who may be called upon at a moment's notice to supply something upon which the lives of some of his fellows and the safety of some valuable piece of work may depend.

Strength of Projecting Veranda. To the craftsman who is desirous of adding a little theory to his practical knowledge, there is nothing, perhaps, so repellant as the apparently difficult arithmetical formulas with which he is confronted in the pages of many of his trade books. The writer's success in making some of these formulas clear to a couple of his craftsmen friends, led to a continuation of the lessons, and he begs to offer an account of the next step taken in helping his two friends past some more little difficulties.

It is a well-known maxim in all teaching, that we "must proceed from the known to the unknown," and it seems that the consideration of the strength of a wooden cantilever beam, just discussed, leads naturally to the question of the method of calculating the strength of verandas and similar structures, which are supported by a number of projecting beams or cantilevers.

The method given by which the strength of the projecting beam used for hoisting goods to an upper story is ascertained, can be easily applied to a series of such beams, and their combined strength readily ascertained.

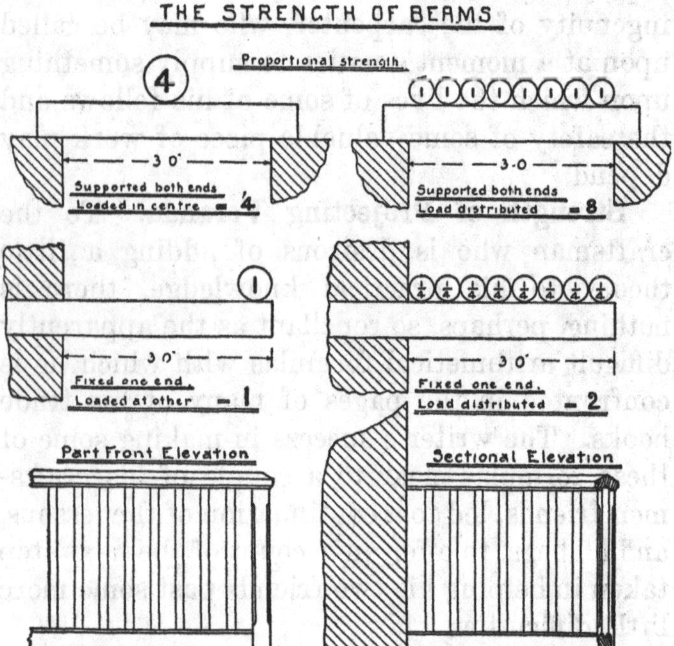

THE STRENGTH OF BEAMS

Fig. 164. **Strength of a Projecting Veranda.**

The first step, then, was to look about for a suitable example of such a structure, the strength of which could be calculated for purposes of the lesson. A very short excursion round the neighborhood led to the discovery of a veranda or balcony which suited the purpose admirably; and our two craftsmen pupils soon measured the structure and jotted down the necessary particulars, which were as follows:

The veranda, Fig. 164, projected from the second story of a dwelling-house some 3 feet, the house itself being 24 feet in width, with the veranda right across the face. It was carried on 11 spruce beams, each 4 by 6 inches, projecting through the face of the wall, giving 10 bays of flooring. There was a sloping roof to the veranda; but, as this was carried on some beams projecting from the floor above, its weight had not to be taken into account. A light balustrade about 2 feet 6 inches high completed the structure.

The first proceeding was to calculate the strength of one beam, or cantilever, of the size used in the veranda. Going back to the first lesson, we proceeded to find the strength of such a piece, supposing it to be a simple beam loaded in the middle and supported at each end, the calculation being as shown in Fig. 165, the steps being as follows:

Put down the figure for spruce (see table) = 3½ cwts. Put down the breadth 4 inches, then the depth 6 inches, and, as that was to be "squared," 6 again. Put underneath, the length, 3 feet. This sum gave us 168 cwt. as the breaking weight when the piece was supported at both ends.

But we had already seen that a cantilever loaded at the end is only one-fourth as strong as the same piece is when supported at both ends and loaded in the middle. In the case of the veranda, however, the load would not be at the end, but would be distributed along the length of the cantilever, thus bringing in another rule, which is: **A cantilever loaded evenly throughout will carry twice as much as a similar one loaded at the end only.** (A similar rule applies when the beam is supported at each end.) Accordingly, we divided our answer by 2, giving us $168 \div 2 = 84$ cwts. as the breaking weight for a spruce cantilever, 4 inches by 6 inches, projecting 3 feet from its support, the wall of the house.

As the veranda would have to carry a live load, we divided this again by 8, the factor of safety, and obtained $84 \div 8 = 10\frac{1}{2}$ cwts. as the safe load for one cantilever.

The next step was to find what safe load the whole veranda would bear. As there were eleven cantilevers, the first thought of the pupils was to multiply the safe load of one of them by 11. A little consideration, however, showed them that it was the number of bays of floor which had to be counted; each bearer having to support half the load of the bays on each side of it. As 11 bearings gave us 10 bays, our sum was 10½ cwts. multiplied by 10, giving us 105 cwts, for the whole veranda.

The next consideration was one about which

THE CALCULATIONS

A
$$\frac{3t \times 4'' \times 6'' \times 6}{3} = 168$$

B
168 cwts. is Breaking Weight of piece when as in Fig. 1.

C
2|168
84 cwts. is B.W. when as in Fig. 4.

D
8|84
10½ c. Safe load (⅛ B. Weight)

E
Verandah has 11 bearer or 10 bays of floor; 10½ cwts. × 10 = 105 cwts.

F
Verandah is 24'. 0" × 3'. 0" = 72 sp.ft.

Load cannot exceed 1½ cwts. per foot.

G
72 × 1½ = 108 cwts., the greatest load possible for whole verandah.

Fig. 165. Steps to be Taken in Figuring Strength and Load of Veranda.

the craftsmen pupils had no definite knowledge
—namely, as to the weight of a number of people standing on a floor. As this is a very useful
thing to remember, it should be noted that
numerous experiments have shown that the
weight of a crowd of people does not exceed one
cwt. per foot super of floor space. "But, suppose," said one of the pupils, "that the veranda
was filled with people excitedly watching a
street procession." ("Or a dog fight," inter-
posed his mate.) "Would not that make some
considerable difference?" As this was a very
proper matter to take into account, it was decided to count upon a live load of 1½ cwts. per
super foot. The veranda being 24 feet by 3 feet
gave us $24 \times 3 = 72$ feet super, which, at 1½ cwts.
per foot, gave us $72 \times 1\frac{1}{2} = 108$ cwts. as the
greatest possible load which would be likely to
be placed on it.

As this came marvelously near to the previously calculated strength of the structure, it
appeared that the designer of it had been fairly
correct, and that it was perfectly safe as long
as its members were not weakened by age, rot,
or other defects.

NOTE—The curved rib or bracket shown under the
cantilever in the drawing, Fig. 164, is almost wholly ornamental and was not considered at all in the calculations.

**To Determine Width of Beam to Support a
Given Load.** "Well, how are you getting along
with your calculations of the strength of timber?" was the writer's greeting as his two

craftsmen friends and pupils came into the office again one day recently.

"Oh! we are doing pretty well, and we can now remember the first rule or formula which you gave us without having to turn to refresh our memories. But we struck a small snag yesterday. We have a sort of solution; but we felt that there was a proper way to work it out, and that is the reason of our visit to-day."

The writer having expressed his readiness to help, the spokesman went on to explain the little problem which was worrying him, somewhat as follows:

"Up to now we have been finding the strength of some particular piece of timber when fixed and loaded in various ways. But suppose we knew the load that was to be carried by a beam, and wanted to calculate the size of the timber—is there not some way of doing that just as easily as finding the strength of a piece whose size we know?"

Having been assured that the calculations for such a problem were only a trifle more difficult than for the one now familiar to them, the speaker went on to state his problem more particularly.

"There is an 8-foot driveway to be made through a new brick store, and we want to find the size of the beam that will safely carry the weight of the brickwork in the story above the beam. We want to have the beam of such a depth that it will be equal to so many courses of the brickwork, and thus save a lot of cutting for the bricklayer and a poor appearance afterwards. The wall is one brick thick (8 inches), and is to be carried up nine feet above the beam. The bricks are 2⅛ inches thick, and four courses measure just 10 inches.''

Having these particulars, we proceeded as follows: First, we found the weight of the brickwork to be carried by the beam. Nine feet high, 8 feet wide, and 8 inches thick, gave us 48 cubic feet. A cubic foot of brickwork weighs about 1 cwt. (112 lbs.); so the weight to be carried was, of course, 48 cwt.

But the roof had also to be considered, and we next proceeded to calculate the weight of that portion of the roof carried on the wall over the beam. From ridge to eaves, the roof slope measured 20 feet; 20 feet times 8 feet gave us 160 square feet. Slates were to be used as coverings; and as the weight of a slate covered roof, allowing for wind pressure, is usually taken at ½ cwt. per square foot, we obtained as the weight to be considered, 80 cwt. (160 times ½).

The problem then resolved itself into this: One hundred and twenty-eight cwt. was to be carried as a **distributed load** on a beam 10 inches deep over an opening of 8 feet span. The load was to be stationary, or "dead." How broad should our beam be to carry the weight safely?

It seemed best to follow the plan adopted in the earlier lessons, and to give another rule or formula which could be easily referred to and

in time remembered. This was written down
as in Fig. 166, which is merely a simple way of
stating a reversal of our first rule for finding the
strength of a given beam.

With this rule before us, we put down our
particulars as in Fig. 167, Calculation No. 1—
namely, 128 cwts. for our load; 8 as our length; 5
as our factor of safety (dead load); 10 times 10
as our depth (squared), equal to four courses of
brickwork; 5 as our figure for oak (see table);
and 2 because our load was to be evenly distributed along the beam.

To find the breadth of a beam when the other data are known

REMEMBER THE RULE.

Put down —

$$\frac{Load \times Length \times Factor\ of\ safety}{Depth \times Depth \times Figure\ for\ wood \times 2} = \begin{array}{l} Result :- \\ Breadth\ in\ inches \end{array}$$

Note The figure 2 is used when the load is evenly distributed
If the load is to be a central one, this figure must be left out

Fig. 166. To Find the Breadth of a Beam Supported at Both Ends
Required for a Given Load Evenly Distributed.

The result, 5⅛ inches, being less than the
thickness of the wall to be carried, led us to try
what breadth our beam would have to be if we
made it only 7½ inches deep—that is, equal to
three courses of brickwork—Calculation No. 2
shows this, which gave as a result rather over
nine inches (see Calculation No. 2, Fig. 167).
As this was more than the thickness of the wall
to be carried, it was decided to make the beam
10 inches deep, and, for the better convenience
of the bricklayer in starting his courses, to make

the beam the full thickness of the wall. This, of
course, was a stronger beam than was actually
required, but was an error in the right direction.

THE STRENGTH OF BEAMS

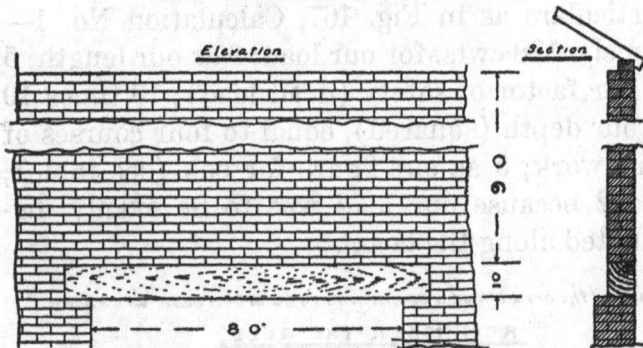

Calculation No 1.

Load on Beam = { Brickwork, 48 cwt / Slate roof 80 " }　Total 128 cwt

Length = 8·0　　Factor of safety = 5　　Depth = 10 in

Figure for wood (Oak) 5　Figure for distributed load　2

$$\frac{128 \times 8 \times 5}{10 \times 10 \times 5 \times 2} = \frac{5120}{1000} \text{ or (about) } 5\frac{1}{8} \text{ ins.}$$

Therefore a piece of oak 10" x 5⅛" is strong enough.

Calculation No. 2

Beam 7½ deep — other particular the same

$$\frac{128 \times 8 \times 5}{7\frac{1}{2} \times 7\frac{1}{2} \times 5 \times 2} = \frac{5120}{562\frac{1}{2}} \text{ or (about) } 9\frac{1}{8} \text{ ins.}$$

Therefore a piece of oak 7½" x 9⅛" (on flat) would also do.

Fig. 167. How to Determine Size of Beam Required above Opening
in Brick Wall.

In point of fact, it is always wise to err on the side of strength in deciding on the sizes of timber to be used in construction, for no two pieces give the same results if tested. As already remarked, the strengths of the various kinds of woods as given in most textbooks were obtained by averaging the results of hundreds of actual tests; and it was found that different pieces, even from the same tree, varied considerably. So serious is this variation, that many architects require that beams or girders carrying heavy weights shall be ripped in half and the ends reversed. It is very common to find in specifications for beams in positions similar to that which is the subject of this present calculation, a requirement that (in addition to the usual clauses as to the quality, etc., of the lumber) the piece be **halved, reversed, and bolted together again.** By doing this, any possible weakness at one end of the stick would be obviated, and a piece of even strength obtained.

Of course, in many parts of the world, iron beams have largely superseded wooden ones; but for a long time to come, wood will hold its own in many districts for a variety of reasons.

To sum up this lesson, then, the first thing is the new rule or formula for putting down the known particulars. In doing this the things to be remembered are: (1) the factor of safety (5 for a dead load, 8 or 10 for a live load); (2) the figure for the particular timber used; and (3) the fact that a beam carries a distributed load just double of what it would carry when loaded in the center only.

To Determine Depth of Beam to Support a Given Load. There has just been given the method of finding the **breadth** of a beam when the **depth** was already fixed, the length (span) and load being also known. The case worked out is only one of numerous instances in which the depth of a beam is limited. A very common case is that of a floor where a beam has to be placed to carry a partition above, but is flush with the lower edges of the joist below. The last formula given will apply to that and all similar cases.

Very often, however, it happens that it is the **breadth** of a beam that is fixed; and the problem then is to find what **depth** the timber should be to carry the load safely. To find this requires, of course, merely a slight turning about of the formulas used in the cases dealt with previously, and presents no difficulty to anyone at all acquainted with mathematics. It will be worked out, however, in the same manner as the other problems, and divested as much as possible of all complicated and repellant algebraic signs and symbols. The writer's attempts to interest and instruct his two craftsmen friends showed that it is usually the look of the formulas in books that scares the seeker after knowledge who is some years away from his schooldays. Also, that if these mystifying formulas are properly and simply explained, any man with a knowledge of the elementary rules of arithmetic can make calculations for the size of practically all the timbers of the average building.

As before, certain conditions will be laid down, being as follows:

A beam is to carry an 8-inch wall over an opening. The beam is of Southern hard (or pitch) pine; the opening is 10 feet wide; the height of the wall above the beam is 13 feet 6 inches, giving 90 cubic feet (10 feet by 13 feet

REMEMBER THE RULE !!!

To find DEPTH of a beam when the following particulars are known:—

(a) LENGTH, (b) LOAD, (c) BREADTH, (e) NATURE OF LOAD,

(f) POSITION OF LOAD, (g) SORT OF WOOD

PUT DOWN —

$$\frac{LENGTH \times LOAD \times FACTOR\ OF\ SAFETY}{2 \times BREADTH \times FIGURE\ FOR\ WOOD} = \frac{Result}{SQUARE\ \text{of the}\ DEPTH}$$

Note The figure 2 must be left out for a central load .

Fig. 168. To Find the Depth of a Beam Supported at Both Ends, Required for a Given Load Evenly Distributed.

6 inches by 8 inches)—say, 90 cwt. as weight to be carried on beam.

The method of putting down these particulars is shown in Fig. 168, which will be seen to resemble the earlier formulas, although the several factors are differently arranged. Fig. 169, **A,** shows the calculations for our present case, which works out in exactly the same manner as earlier ones; that is, all the values **above** the line are multiplied together, and divided by the product of all the values **below** the line.

The only difficulty is in working out the final answer, for the result obtained at first will be the **square of the required depth**—that is, the depth multiplied by itself. To arrive at the exact depth, it is necessary to **extract the square root** of the answer, which means to find what number multiplied by itself will give the answer. For all ordinary practical purposes of wooden beam calculations, however, the exact result is not absolutely necessary, and a sufficiently accurate one can be obtained by inspection of the first answer. For instance, in the foregoing problem the answer is $56\frac{1}{4}$. Now, the nearest square of a whole number to this is 49, the square of 7 ($7\times7=49$); therefore the depth of the beam is more than 7. The next square of a whole number is 64, the square of 8 ($8\times8=64$); and, as that is greater than our answer, evidently our beam should be somewhere between 7 and 8 inches deep. As a matter of fact, it proves in this case to be exactly $7\frac{1}{2}$ inches ($7\frac{1}{2}\times7\frac{1}{2}=56\frac{1}{4}$); but if the sum had not worked out so exactly as that, a result quite good enough for practical purposes could have been arrived at by the method indicated above— namely, by finding the nearest squares of the whole numbers above and below the answer, and allowing a sufficient amount over and above the root number of the square below. As in the previous articles, the nature and position of the load must be considered in any calculations made. The factor of safety used is 5, the usual for a dead load. That is, one-fifth of the breaking weight is considered to be the amount a beam can safely carry when the load is a stationary one.

The position of the load is important, for, as shown in the earlier lessons, a beam will carry twice as much if the load is evenly distributed along it, as it would if the load were in the cen-

A. _For distributed load_

$$\frac{10 \, ft. \times 90 \, cwt. \times 5, \, factor \, for \, dead \, load}{2 \times 8in \times 5, \, figure \, for \, pitch \, pine} = \frac{4500}{80} = 56\frac{1}{4}$$

The square root of $56\frac{1}{4}$ is $7\frac{1}{2}$, the required depth for beam

B. _For central load_

$$\frac{10 \times 90 \times 5}{8 \times 5} \quad 112\frac{1}{2} \qquad$$ The square root of $112\frac{1}{2}$ is nearly $10\frac{1}{2}$

A piece 8"x7½" would carry the given load if it were evenly distributed

A piece 8"x10½" would be required if load were central

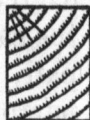

Fig. 169. Calculation for Finding Depth of Beam to Support a Given Load.

ter only. The figure 2 is therefore placed below the line in the present case. The effect of this is shown by comparing the result at **A** with that at **B**, Fig. 169.

From the several calculations made, it will be seen that if two beams or joists are of the same length and sectional area, the one of greater depth will be the stronger of the two.

This can readily be seen by taking as examples two joists of the same sectional area, but of different dimensions. A piece of 12-inch by 2½-inch, another 10 inches by 3 inches, have the same sectional area (30 square inches); but their relative strength when placed on edge is as 360 to 300. The rule stated covering this, is that **the strength of a beam is as the square of its depth**—that is, the depth must be multiplied by itself; $12 \times 12 \times 2\frac{1}{2} = 360$, and $10 \times 10 \times 3 = 300$, or a proportion of strength between the two as 6 is to 5.

There is, of course, a limitation in the practical application of this; for if a beam be made very deep in relation to its breadth, it will buckle and twist when loaded. In the case of floor-joists, the disproportion of depth to breadth is very marked; but their tendency to buckle is overcome by strutting, either with solid blocks the same depth as the joists cut in between each pair, or with herring-bone strutting cut from narrow battens.

Many experiments have been made to find the best proportion for the breadth and depth of wood beams, and it has been laid down that a ratio of 5 to 7 gives the best section. This is a useful thing to remember and easily kept in mind.

In concluding our discussion of the strength of beams, the writer trusts that some of the readers who may have been deterred from going into the matter of calculating the strength of materials will be in some measure led to see that a formula is only a simple way of putting down a rule for some arithmetical process that would take a long time to describe in words. The strength of beams is a question that is so often cropping up that this discussion may be directly valuable in showing how to find it in any given case; but the writer also hopes that the discussion may lead many readers to take up other lines of calculation equally simple and useful to the practical man.

Truss Construction

The principle of a **truss** is theoretically a number of straight bars joined near their ends by flexible joints, and arranged so that all their internal stresses are sustained by its members, and only the vertical pressures (the weights of the truss and its load) are transmitted to its abutments. Trusses differ from **solid beams** inasmuch as the weight of the truss and its load may be regarded as divided into portions which are concentrated at the joints between the members, and which act through the centers of gravity of their cross-sections. So placed, the stresses caused by them could not act **transversely** of the members, as in a beam, causing secondary stresses, but must act **longitudinally** of the members, and must be uniformly distributed over their entire cross-sectional areas. This is the distinguishing feature of all trusses; while in a solid beam, when it bends under its load or its own weight, all the fibers above the neutral axis are compressed, and all those below are extended, the resulting change of length in each fiber being proportional to the distance of the fiber from the neutral axis.

Most of the trusses in common use consist of two long members, called **chords,** extending the entire length of the span, and connected by

web members, which are sometimes all inclined
and sometimes alternately vertical and inclined.
Inclined web members are called **diagonals**, such
web members being known as **ties** and **struts**.
A member sustaining tension is called a **rod** or
tie; and one sustaining compression is called a
strut or **post**; while one capable of sustaining
both tension and compression is called a **tie
strut**.

The simplest form of truss consists of a single triangle (Fig. 170). Truss **a** is in com-

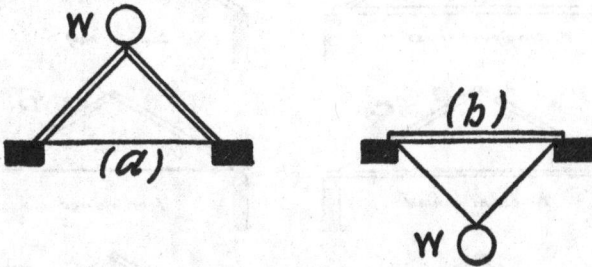

Fig. 170. Simplest Form of Truss.

pression in the rafters, and in tension in the
chord or tie-rod; and truss **b** is in compression
in the chord, and in tension in the tie-rods—the
reverse of **a**. This, of course, is in common use
for roofs of small span, as in dwellings, and, in
practice, is loaded along the rafter, and not
alone at the apex as in **a**; but in calculating
the stresses in the members, we commonly first
assume that the loads are concentrated at the
intersection of the truss members, and the effect
of actual distribution along the members is then
determined, separately treating the members as

beams. Other more elaborate and complicated
trusses are all built up on this principle of the
simple triangle.

Simple Roof Trusses. A simple structure
which is sometimes misunderstood is a common
king-post roof truss, such as is shown at **A** in
Fig. 171.

If two rafters are placed as at **B**, they will,

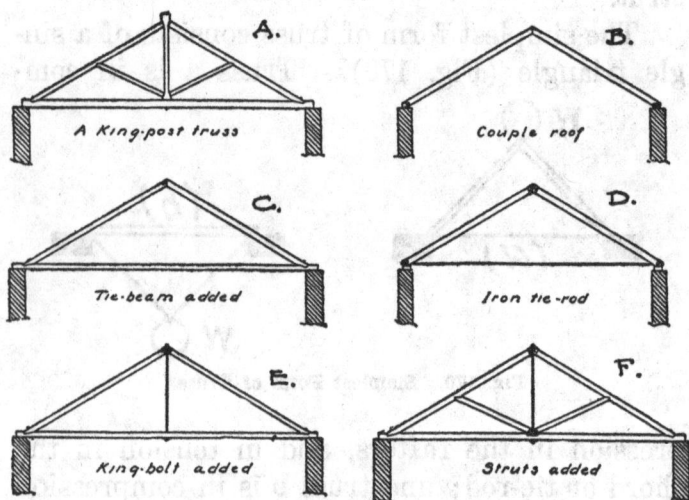

Fig. 171. Simple Roof Trusses.

of course, tend to spread at the bottom and push
the walls apart. This is prevented by a wooden
tie-beam, as at **C**, or by an iron tie-rod, as at
D. But after a certain span has been exceeded,
the tie-beam commences to sag and has to be
supported either by a king-post or by a bolt
from the apex of the rafters, as at **A** and **E**.

The writer has found many carpenters with

the impression that the king-post rests on the
tie-beam to support the ridge; but a little reflec-
tion and a study of these diagrams should con-
vince even the beginner in roof framing of the
fallacy of that view. As a matter of fact, both
the tie-beam and king-post are in tension; that
is, they are being pulled, and their work can
be, and often is, done equally well with a
wrought-iron rod.

But, as the span becomes so great as to necessitate the support of the tie-beam from above,
the rafters will also become so long as to tend
to sag in the center, and they in turn must be
supported.   This is done by struts from the
foot of the king-post or rod, as at **F**. It is clear
that these struts are in compression, and that
they must be of wood or some other stiff
material, such as angle or T-iron, in the case of
an iron roof or other roof of heavy construction.

There are thus two kinds of stresses in such
a truss, the tie-beam and king-post being pulled,
or **in tension**; and the struts and rafters being
pressed, or **in compression**. For the former, a
flexible material, such as a wrought-iron rod, is
suitable; but for the latter, a stiff and unyield-
ing material must be employed, so as to resist
the tendency to buckle or bend under the weight
of the roof.

Light Trusses for Broad Span. Two light
timber trusses designed for broad span are
shown in detail in Figs. 172 and 173. They are
suitable to support the roof of a hall or rink.

They are designed for a span about 65 feet wide,
and are intended to carry nothing but the roof.

The top and bottom chords, Fig. 172, will

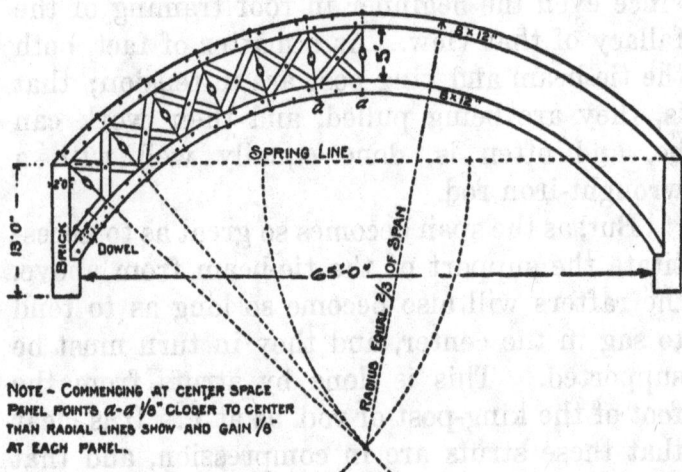

NOTE – COMMENCING AT CENTER SPACE
PANEL POINTS *a-a* ⅛" CLOSER TO CENTER
THAN RADIAL LINES SHOW AND GAIN ⅛"
AT EACH PANEL.

Fig. 172. Strong, Curved Roof Truss.

have to be built up segmental. Use ½ by 4-
inch iron jib strap on bolts, so as to catch all
the timbers. Braces and counters can be

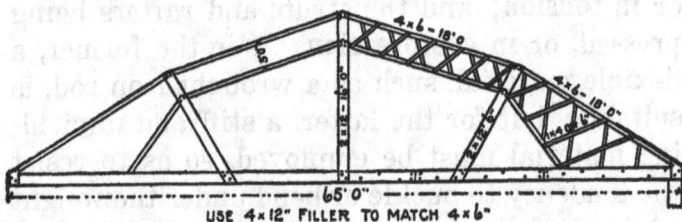

USE 4×12" FILLER TO MATCH 4×6"

Fig. 173. Cheap Roof Truss for Broad Span.

dressed, and chord left rough and boxed afterward. Have all rods fitted with turnbuckle in
the center, as the end will not be exposed. Have

tension on all rods as near the same as possible. Make first panel point ⅛ inch closer than radial lines shown, which will make the next one ¼, and next ⅜, and so on, to allow for compression in the top chord. Brace bottom transversely to prevent warping.

If it is not desired to have finished appearance, a cheaper truss for the same use may be made as shown in Fig. 173.

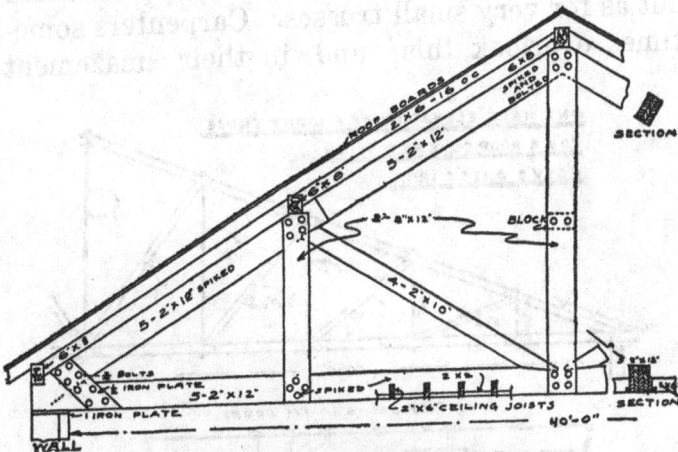

Fig. 174. Cheap Plank-Framed Truss.

Plank-Framed Truss. A plank-framed truss is very popular where a cheap truss is wanted to support a roof of say 40-foot span, as for a hall. Fig. 174 shows a well-designed built-up truss of satisfactory and at the same time cheap construction, for such a purpose. The wind and snow load on a roof of this kind is a factor that has to be considered. Such trusses will prove amply safe for a 40-foot span. They should be set in 16-foot bents. The expensive large-dimension timbers are not used, the different members being built up of 2 by 10 and 2 by 12-inch pieces, spiked together to break joints as specified on the drawing.

Light Truss for 100-Foot Span. In designing trusses of wide span such as those used to support temporary roofs of light construction, the same principle of triangulation is carried out as for very small trusses. Carpenters sometimes overlook this; and, in their amazement

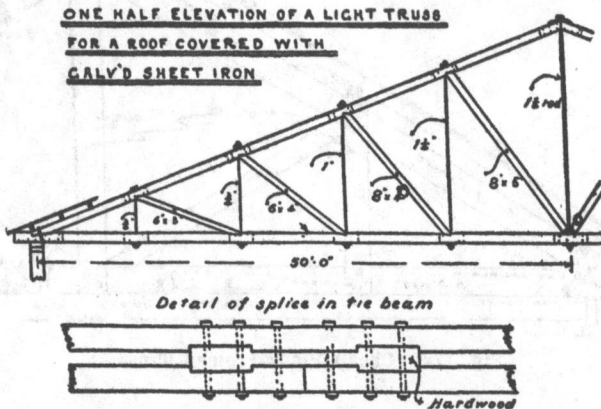

ONE HALF ELEVATION OF A LIGHT TRUSS
FOR A ROOF COVERED WITH
GALV'D SHEET IRON

Detail of splice in tie beam

Fig. 175. Light Truss for 100-Foot Span.

at the width of span, trying to arrange framing adequate to the task, lay out a series of members not a truss at all, but a number of quadrilateral panels which offer little resistance to change of form, and which would easily be racked by an extra force acting on one side, such as a gale of wind.

Fig. 175 shows a good form for a long-span truss. For the bottom chord (tie-beam), two pieces of 9-inch by 4-inch, at least, will be necessary. The splicing of the pieces to obtain length enough for the span will reduce the sectional area considerably, and must be allowed for. The system of tension rods and struts must be carried out carefully as to joints. The blades or rafters (top chords) should be of two pieces of 8-inch by 3-inch, blocked and bolted together at frequent intervals. The sizes of the other members are shown in the diagram. It is very much better to use iron rods for all the tension members as shown. If wood is used, straps and bolts at top and bottom of each member will be required, to hold up the weight of tie-beam and thrust of strut. If the rods are upset at the ends, and a plus thread cut upon them, slightly smaller iron can be used.

It is usual to allow on trusses of this description a camber of half an inch for every ten feet of span. Five inches may seem a lot for this truss, but is none too much when the number of joints is considered.

A Lattice Truss. A cheap truss for broad spans is the **lattice truss**, built up out of light timbers which can be had in any lumber yard. It is easily constructed. Fig. 176 shows such a truss of 60-foot span. No unsightly rods are required to keep the side walls from spreading, since in this form of truss, the truss itself acts as a tie, and, when properly anchored, there can be no tendency to crowd the walls out. Six trusses, exclusive of the ends, will be sufficient for a building 80 feet long, which would call for a spacing a little less than 12 feet from centers. The covering for this form of roof may be of almost any of the roofing materials, aside from the gravel-coated, as the inclines at the sides are too steep to stand for any great length of time the wind and wash that the gravel roof would be subjected to.

Trusses for Flat Roof. A carpenter recently had a typical flat-roof store building to erect.

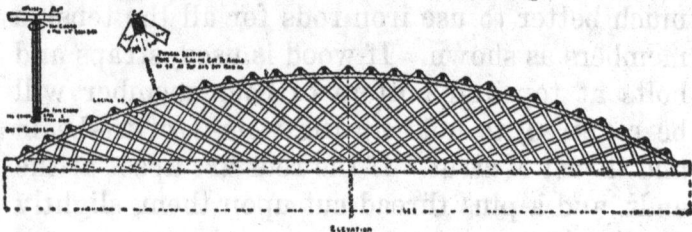

Fig. 176. Cheap Lattice Truss.

It was to be 44 feet wide and 100 feet long, the roof to slope from front to rear and to be covered with gravel roofing. The building was two stories high, one room on each floor. The following advice was asked: "How would you frame roof without having columns under second-story ceiling? Second floor is to be used for skating rink. How would you deaden floor? Joists are to be 2 by 12-inch hard pine."

The best way to carry such a roof is by a series of trusses, as shown in Fig. 177. The heights of these trusses vary according to their position, a fall of 1 inch to the foot being allowed for the slope of the roof. The truss shown in elevation in Fig. 178 is the lowest of the series; but the rest would be similar in all respects.

It is difficult to deaden a floor effectively when used for such a noisy purpose as roller

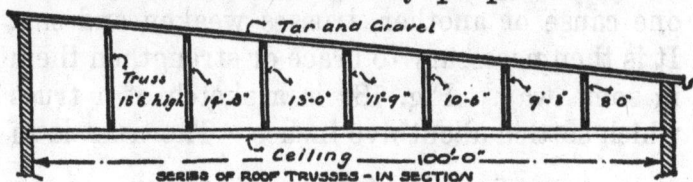

Fig. 177. How to Support a Flat Roof.

skating; but we advise a double deadening as most likely to fill the bill. One method, shown in Fig. 179, consists in nailing strips near the lower edges of joists to support short lengths of boards, upon which is laid a rough mortar either of sand and lime, ashes and lime, or sawdust

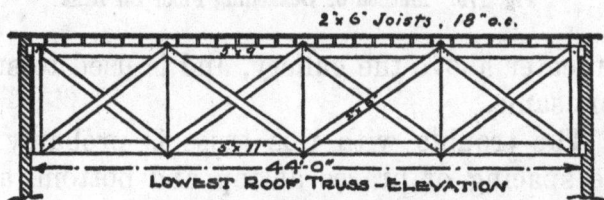

Fig. 178. Truss for Flat Roof.

and lime. Or one of the special patent compositions, such as slag wool, made for the purpose, might be used.

The second deadening is applied between the rough floor and the hardwood upper flooring, and consists of heavy deadening felt made for

the purpose.　In the case of a rink, the felt
should be laid on the rough floor; then strips
of 1 by 2-inch stuff laid flat, 12 inches apart; and
then the hardwood floor laid to form the skating
surface.

To Strengthen a Truss.　Occasionally, from
one cause or another, trusses weaken and sag.
It is then necessary to brace or strengthen them
in some way.　Fig. 180 is a sketch of a truss
which settled about five inches.　The truss is in

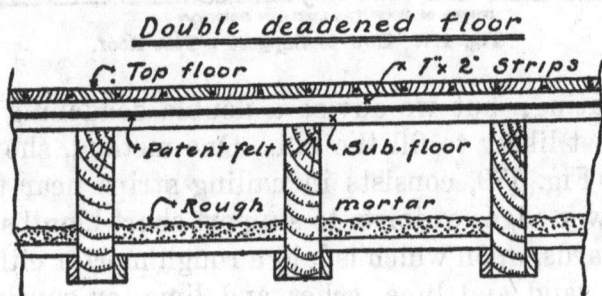

Fig. 179.　Method of Deadening Floor for Rink.

a theater above the gallery, and is used to support same.

The trouble with this truss is probably in
the spacing of braces at top and bottom, and
changing them would be impracticable.　To
make it perfectly safe, it would do to jack it up in
center until it has sufficient camber, and then
reinforce the bottom chord with a couple of 2
by 12-inch planks, one bolted on each side, and
put a 1¼-inch truss-rod on each side, as shown
in sketch, with turnbuckle.　If ends of rods are
not upset, to depth of thread, use heavier ones,

so that the diameter at bottom of thread will be
not less than 1¼ inches. This will result in a
change of length of braces, and they should be
replaced. As the middle vertical rod carries no
part of the load of the truss proper, it is not
necessary to have nut and washer under bottom
chord; and the rod may pass on down through
the gallery. Theoretically the rod serves only
to prevent deflection in the two unloaded middle
chord panels under their own weight, but in
practice it is usually employed for convenience.

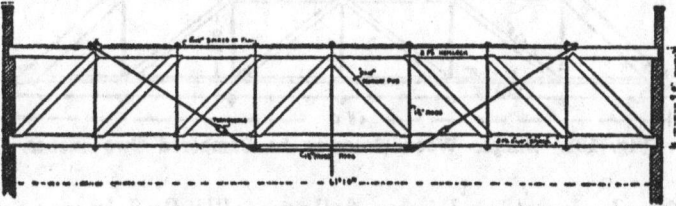

Fig. 180.   Strengthening a Truss with Rods.

Top chord should be reinforced with a 2 by 10
spiked on flat, between angle washers. See that
foot of end brace is held rigidly in place.

To Camber a Truss. Fig. 181 is a truss,
specification for which says "frame it so it will
have 2-inch camber when tightened up." Now,
the question is, how to get the proper location of
the holes for the rods, also the lengths of each
stick of timber.

The method of cambering such a truss in-
volves some knowledge of mathematics and
drawing. Nearly all engineers are now in the
habit of making the necessary calculations and
figuring the exact lengths of the braces upon

the drawings before they are sent out. The proper angle of the cast-iron shoes is also worked out, and a full-sized drawing of each casting is made in the drafting room.

In theory, when such a truss is cambered, the upper chord becomes longer than the bottom. The panels will thus be out of square and the braces slightly longer in consequence. A rule for finding the increase of length has been

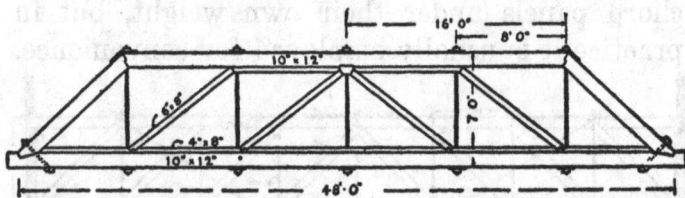

Fig. 181. Simple Wood Truss to be Cambered Two Inches.

worked out and is as follows: To find increase in length of upper chord, put down:

$$\frac{\text{Depth of truss} \times \text{Camber} \times 8}{\text{Span}};$$

(all in feet or inches). Applying this rule to the problem at hand, we have:

$$\frac{7 \text{ ft.} \times 1\text{-}6 \text{ ft.} \times 8}{48} = \tfrac{7}{8} \text{ ft., or } 2\tfrac{1}{3} \text{ inches.}$$

The figure 8 in the formula is a constant, and is used in all cases.

The increase found by working this very simple calculation, is divided amongst the panels. If your drawings have not been figured for camber, you will require to make a full-

sized drawing of one panel of the truss upon a
board platform or convenient floor. The drawing should show the panel as much wider at the
top as your calculations will direct, and the
braces can then be cut to length and bevel on
your drawing. Of course, the upper chord is
not actually lengthened $2\frac{1}{3}$ inches, all that is
necessary being to cut your braces to fit the
full-sized drawing of the distorted panel. In
this case the distortion (out of square) is very
small, as the truss is shallow. It may be taken
as one-half inch for each panel; and the braces
may be made five-eighths or three-quarters of
an inch longer than they would be if panel were
square.

This applies, of course, not to all cases, but
only to the case of a simple wooden truss. It
must be noted that in steel trusses, where the
braces abut against machined surfaces, very
exact calculations are necessary for finding the
lengths of braces, and angles of bearing surfaces. In large wooden trusses, great care is
also taken in this respect, although it is easier
to adjust the length and cuts of the braces in
this material.

The positions of bolts are obtained by
spacing evenly, as shown, and should present no
difficulty.

Trussed Partitions. In a four-story building, ground plan 42 by 72 feet, the first floor
was to be one large room without posts. Two
trusses on the second floor, each in a partition

running across the building, were desired. The ceiling on the second floor was ten feet. The diagrams, Fig. 182, show two trussed partitions, 42 feet by 10 feet, with the doors arranged as desired. The studding is omitted for the sake of clearness, but would be cut in between the

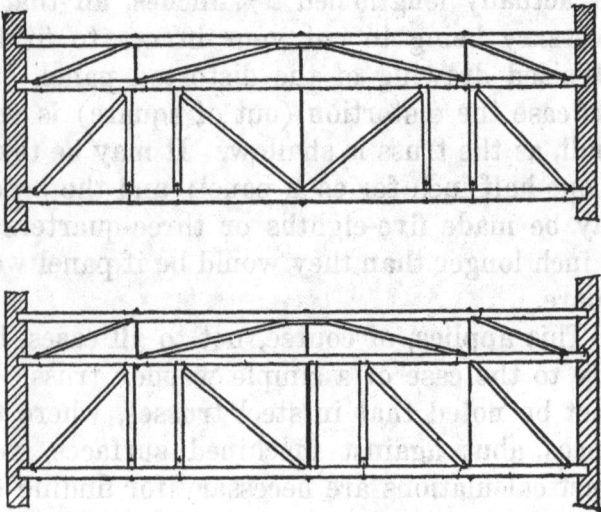

Fig. 182. Trussed Partitions.

ties and braces and spiked to them at proper distances apart.

If these trussed partitions were calculated to carry simply their own weight, the dimensions of the framing would be as follows: Bottom member, 10 inches by 5 inches; inter-tie and top member, 8 inches by 5 inches; long struts, 4 inches by 5 inches; upright members, 4 inches by 5 inches; tension rods, 1¼-inch iron; other bolts, ⅝ inch.

Index

335

338 INDEX

www.ingramcontent.com/pod-product-compliance
Lightning Source LLC
Chambersburg PA
CBHW011225210326
41598CB00039B/7312